¡ALERTA!

Engineering Studies

Edited by Gary Downey and Matthew Wisnioski

Matthew Wisnioski, *Engineers for Change: Competing Visions of Technology in 1960s America*

Amy Sue Bix, *Girls Coming to Tech! A History of American Engineering Education for Women*

Jessica Smith, *Extracting Accountability: Engineers and Corporate Social Responsibility*

Elizabeth Reddy, *¡Alerta! Engineering on Shaky Ground*

¡ALERTA!

ENGINEERING ON SHAKY GROUND

ELIZABETH REDDY

The MIT Press
Cambridge, Massachusetts
London, England

The MIT Press would like to thank the anonymous peer reviewers who provided comments on drafts of this book. The generous work of academic experts is essential for establishing the authority and quality of our publications. We acknowledge with gratitude the contributions of these otherwise uncredited readers.

This book was set in Stone Serif and Stone Sans by Westchester Publishing Services. Printed and bound in the United States of America.

Library of Congress Cataloging-in-Publication Data

Names: Reddy, Elizabeth, author.
Title: ¡Alerta! : engineering on shaky ground / Elizabeth Reddy.
Description: Cambridge, Massachusetts : The MIT Press, [2023] |
    Series: Engineering studies | Includes bibliographical references
    and index.
Identifiers: LCCN 2022029552 (print) | LCCN 2022029553 (ebook) |
    ISBN 9780262545518 (paperback) | ISBN 9780262374378 (epub) |
    ISBN 9780262374385 (pdf)
Subjects: LCSH: Earthquake prediction—Mexico—History. | Environmental
    monitoring—Mexico—History.
Classification: LCC QE538.8 .R43 2023 (print) | LCC QE538.8 (ebook) |
    DDC 551.220972—dc23/eng20221028
LC record available at https://lccn.loc.gov/2022029552
LC ebook record available at https://lccn.loc.gov/2022029553

10 9 8 7 6 5 4 3 2 1

CONTENTS

SERIES FOREWORD

We live in highly engineered worlds. Engineers play crucial roles in the normative direction of localized knowledge and social orders. The Engineering Studies Series highlights the growing need to understand the situated commitments and practices of engineers and engineering. It asks: What is engineering for? What are engineers for?

Drawing from a diverse arena of research, teaching, and outreach, engineering studies raises awareness of how engineers imagine themselves in service to humanity, and how their service ideals impact the defining and solving of problems with multiple ends and variable consequences. It does so by examining relationships among technical and nontechnical dimensions, and how these relationships change over time and from place to place. Its researchers often are critical participants in the practices they study.

The Engineering Studies Series publishes research in historical, social, cultural, political, philosophical, rhetorical, and organizational studies of engineers and engineering, paying particular attention to normative directionality in engineering epistemologies, practices, identities, and outcomes. Areas of concern include engineering formation, engineering work, engineering design, equity in engineering (gender, racial, ethnic, class, geopolitical), and engineering service to society.

The Engineering Studies Series thus pursues three related missions: (1) advance understanding of engineers, engineering, and outcomes of engineering work; (2) help build and serve communities of researchers and learners in engineering studies; and (3) link scholarly work in engineering studies to broader discussions and debates about engineering education, research, practice, policy, and representation.

Gary Downey, Editor

ACKNOWLEDGMENTS

This book was a collective effort: not just in its research and analysis, but in writing, rethinking, and refining. It has the shape it does because of the ways that people have challenged me, invited me into their projects, and allowed me to find alignments and contradictions within our mutual interests. The recitation that follows is incomplete.

I am grateful, first, for the support of many people hard at work on risk mitigation technologies. I thank those at CIRES, particularly Director Juan Manuel Espinosa Aranda. Armando Cuéllar García, Alejandro Jiménez, Antonio Uribe Carbajal, Oscar Huerta Martínez, Arminda Rangal López, Roberto Islas Vázquez, Armando García, Guadelupe Rico Zetina, Cecilia Hernández Meza, Raquel Macías Troncoso, Sofia Huerta Lledias, Sandra Ramos Pérez, and Lucio Camarillo Barranco. At CENAPRED, Tomás Sánchez Pérez and Hussein Priego Martínez have allowed me access to documents and thoughtful conversation. Roberto Arturo Alvarado at Centro Historico was an informed guide in physical space, as Anders Meira has been in the socialities of business. Fernando Heredia Zavoni at Mexican Society for Seismic Engineering and Heidi Tremayne at the Earthquake Engineering Research Institute opened doors. Elia Arjonilla Cuenca's broad experience and kind, careful reflections have been especially crucial for my insights into Mexican risk mitigation.

I am pleased to be part of ongoing conversations with the ShakeAlert Social Science Working Group team, directed by Sara McBride and Bob DeGroot, who have helped me see earthquake early warning in new ways.

The engineering studies scholars I work with at Colorado School of Mines support and challenge me. How incredible to think and work with these people. This book was shaped by Jessica Smith, Dean Nieusma, Juan Lucena, Marie Stettler Kleine, and Qin Zhu. All of them continue to teach me how to be a scholar, collaborator, and teacher. I am grateful to them and to the International Network for Engineering Studies which brings us (and other like-minded researchers) together.

I also owe a great deal to people who have invited me into new spaces and helped me to make my intellectual and practical projects bigger than I could have imagined alone. Key among these: Michelle Camacho, Susan Lord, Gordon Hoople, Austin Choi-Fitzpatrick, Alex Mejia, and Bre Przestrzelski, who I met at University of San Diego; and Dave Wald and Sabine Loos at USGS. Educators, scholars, and administrators at Mines have picked up where they left off. These include: Alina Handorean, Chelsea Salinas, Mirna Matjik, CJ McClelland, Mark Orrs, Kate Youmans, Leslie Light, Yosef Allam, Kevin Moore, Shannon Mancus, Julia Roos, Jeff Shragge, Tom Williams, Qin Zhu, Rocky Clancy, Nicole Smith, Monica Kurtz, and Kim Walker. I have appreciated teaching and learning from Lanie Breckenridge, Allison Palmer, Julianna Valenzuela, Cassidy Grady, Nick Yavorsky, Leiaka Welcome, Sarah de St. Aubin, Nina Guizzetti, Saad Elbeleidy, and Max Silver. The motivation, creativity, and thoughtfulness that they demonstrate is inspirational.

This project was incubated at the UC Irvine Anthropology Department and through the California STS Network. Here, the ideas in this book benefited from the guidance of great mentors. Bill Maurer did what I understand to be his usual magic. Work with him, Valerie Olson, Kavita Philip, Mei Zhan, and Julia Elyachar kicked things off. I am grateful to Edna Suárez-Díaz and Gisela Mateos González for allowing me to join their excellent UNAM seminar on the history of science. I thank the team at the Institute for Money, Technology and Financial Inclusion, especially Jenny Fan, for giving me a vision for what good institutional and scholarly action can be. I learned a great deal with and from Chima Anyadike-Danes, Georgia Hartman, Colin Cahill, Ben Cox, Sean Mallin, Caitlin Fouratt, Natali Valdez, Simone Popperl, Leah Zani, Alexandra Lippman, Robert Kett, Leksa Lee, Janny Li, Nandita Badami, Taylor Nelms, Lilly Irani, Ellie Harmon, and Nick Seaver.

This book was shaped by other mentors, too. Sandra González-Santos has asked the best and most terrifying questions. Megan Finn's scholarship and critique have been exhilarating. Gwen Ottinger, Jessica Smith, Caitlin Wylie, and Christy Spackman have reminded me how to do what needs doing and avoid what doesn't. Jeremy Spoon does his sophisticated anthropology practically. Charis Boke has demonstrated care for concepts, humans, and nonhumans. Emily Brooks has unsettled my ideas about critical scholarship in practice.

The Committee for the Anthropology of Science, Technology and Computing and the General Anthropology Division of the American Anthropological Association have helped me see the best parts of anthropology. Jenny Cool, Kwame Harrison, Bob Myers, Jenny Carlson, Baird Campbell, Angela VandenBroek, and Rebekah Cupitt have been especially wonderful to work and think with.

Generous writing companions have helped along the way. These include Hayden Kantor, Anaar Desai-Stephens, Ashley Smith, Lara Houston, Vincent Ilianti, and Stephanie Steinhardt. Malte Zweitz is a great host and masterful conversation conductor.

The growing STS as Critical Pedagogy and CREATE/STS communities give me a place to think and play. I salute masterminds Emily York and Shannon Conley who make both possible, as well as the storyteller and scholar Marisa Brandt and new inspirations David Tomblin and Nicole Mogul.

Thanks to Bucknell University Department of Sociology and Anthropology and Field Research and Teaching Lab, The Shiley-Marcos School of Engineering at the University of San Diego, and the Departments of Engineering, Design, and Society and Geophysics at Colorado School of Mines for hosting me (and various incarnations of this book). Research funding for this project was provided by University of California at Irvine's School of Social Science, the University of California at Irvine's Department of Anthropology, the American Institute of Physics, the Society for the History of Technology, the University of California Institute for Mexico and the United States, the National Science Foundation (Doctoral Dissertation Research Improvement Grant #1357388), and the Newkirk Center for Science and Society.

Various chapters of this book have benefited from attention in other spaces. Material has been presented at conferences, workshops, and in talks at Iberoamerican University, Bucknell University, Drexel University, Arizona State University, Portland State University, UC Berkeley, the Catholic University of Chile, and Goldsmiths University of London. Portions of chapters, in significantly different form, have been published in *Ethnos*, *Journal of Political Ecology*, and *Revista Iberoamericana de Comunicación*. I am grateful for comments from listeners, participants, reviewers, and editors that the work has accumulated along the way. I celebrate the labor of development editors Audra Wolfe and Beth Sherouse, who put the manuscript in order. Engineering Studies Series Editors Matt Wisnioski and Gary Downey helped this book find its shape. The MIT Acquisitions team of Katie Helke and Laura Keeler guided me through, and anonymous reviewers imagined what it could be.

Additional friends and relations have made me welcome in their homes and lives as I have researched and written, among them Jennifer Henry, Colleen Duffy, Peter McMahan, Amanda Cort, Elliot Levin, Lauren Mitchell, Sabrina Gogol, Vanessa Gennarelli, Rosamund and Fred Rea, and Kinley and Bonnie Reddy. Some will be able to see their influence in this document more clearly than others will, though it is surely there. Wyatt saw the book through from first inkling to first real full draft, and Olive, ARC, and Ferdinand supported the rest of the process in their own ways. My thanks finally and fully go to Stephen Campbell Rea. SCR, I'm so happy you made a home with me.

INTRODUCTION

ENVIRONMENTAL MONITORING ON SHAKY GROUND

A terrible earthquake shook central Mexico on September 19, 2017. I was not in Mexico then, but I could not stop reading about what had happened. From my office in San Diego, I paged through news reports about the earth moving and the damage it caused. The videos I watched showed sickly swaying buildings, rising dust clouds, and crowds of people seeking safety. My research on earthquakes and earthquake risk mitigation meant none of it was exactly unexpected, but that did not make it easy to take in. I had lived in Mexico City on and off for several years. My feeling for the city and for people living there made the earthquake hard. It made me even more determined to write this book.

Desire to reduce the harm that earthquakes cause is nothing new, but novel technologies are increasingly involved in these projects. Data generated by seismometers and accelerometers have allowed twentieth- and twenty-first-century researchers to collect masses of information about earth motion and put it to uses unimaginable in previous environmental monitoring regimes. Technoscientists can now render the earth as a place constantly moving and crisscrossed by faults and tectonic plate interfaces. They can study the deep composition of the planet and detect secret bomb tests. The proliferation of seismic monitors makes new strategies for risk mitigation imaginable. One such strategy, developed in famously seismic Mexico, involves the Sistema de Alerta Sísmica Mexicano (SASMEX).

SASMEX relies on a network of accelerometers to automatically register quakes as they start and send out warnings. Mexico City can sometimes get more than a minute of notice before shaking that starts on the Pacific coast of Mexico reaches them, and SASMEX can offer at least ten or twenty seconds of warning to places elsewhere in Mexico, too. It may not be very much time, but it is better than nothing. When SASMEX went online in 1991 to provide public earthquake early warnings, it was the first system of its kind in the world. At that point SASMEX used just twelve stations strung across Mexico's western coast. Since then, maintained and championed by a small community of Mexican engineers, this state-funded system has expanded to include ninety-eight seismic field stations and provide alerts to six cities. Proponents suggest that this system, and others like it that have emerged around the world, can help ordinary people and automated systems alike prepare for oncoming earthquakes in ways that will save lives and limit economic losses. They think these technologies can change life with earthquakes.

Earthquakes are still, to the best technoscientific knowledge, unpredictable.[1] Advocates of SASMEX hope that having a warning a few seconds before an earthquake can make a big difference for protective action to save lives or prevent damage. While vocal critics remain, the promise of a chance to take shelter, protect crucial infrastructure, and shut down automatic processes when strong earthquakes are on their way has rallied a great deal of attention. The growth of the Mexican system and the proliferation of similar technologies around the world demonstrate the promise that governments, industries, and ordinary people see in earthquake early warning. This strategy for risk mitigation is important not only because of its growing popularity. It is also part of a broader trend in which technoscientific environmental monitoring tools are made increasingly crucial to our relation to the planet we live on.[2] Because of this, considering SASMEX's development and use in Mexico can help us understand other attempts to engineer new relationships between threatening environments and societies at risk, too.

In this book, I tell the story of a Mexican earthquake risk mitigation technology to demystify such projects. Engineers at the nonprofit Center for Seismic Instrumentation and Registry (Centro de Instrumentación y Registro Sísmico in Spanish, more commonly referred to as CIRES) have designed, developed, maintained, and advocated for SASMEX in the context of changing environmental and social circumstances. Antonio Duran, one of the engineers at CIRES, told me that in any problem with earthquakes, you have to deal with nature, humans, and technology. "You have to put them on the map together," he said.[3] At CIRES, I saw Duran and his colleagues building that "map" as they went. While each element at stake here—"nature, humans, and technology," as Duran put it—may seem easy to define and conceptualize separately, SASMEX demonstrates just how inextricably linked and, indeed, co-constitutive they are. Here, I make the case that the work Duran was describing at CIRES was about rearranging "nature, humans, and technology" on that metaphorical map, or, as I put it throughout this book, transforming relationships between environment, society, and technology.

As globally relevant as they are, both this technology and the particular notions about what it might do do—what historian Matthew Wisnioski might call its "ideology of technological change" —came to be in the context of specific conditions which I seek to address in this book.[4] Public earthquake early warning was first developed in Mexico. I point this out here because that context has shaped its conditions of possibility. It is especially worth emphasizing when technologies in Latin America, no matter how innovative or consequential they may be, are still under-considered in English-language science and technology studies (STS) literature. There is substantial reason to understand that what we often call European or Western technoscience is built with ideas and materials from colonized places generally considered peripheral to the formation of such knowledge. Knowledge production and innovation in such places was nonetheless substantial, and has never stopped.[5] Despite these fact, popular narratives still treat technoscience in Latin America as "imported magic."[6] Unfortunately, we who read in English are still more likely to encounter analyses about technoscience in the United States or Europe than in Latin America. This is a tremendous loss for those of us who seek to better understand technologies and how they are developed, redeveloped, adapted, and used.

Throughout this book, you will find descriptions of Mexican technology grounded in Mexican material environments and sociopolitical realities. This book is, after all, the product of research on earthquake early warning in one very specific place where technoscientific innovation, limited and inequitably distributed wealth, and traumatic seismic histories are all undeniably present. However, I take every opportunity to suggest how Mexican experiences, and the insights of Mexican engineers, can illuminate puzzles and help us better understand the challenges of environmental monitoring and risk mitigation elsewhere, too. In the chapters that follow, I develop three major arguments to help demonstrate what matters in efforts to change how people live with threatening environments.

First, I argue that the development and use of any risk mitigation technology that seeks to remake human–environment relations is a sociotechnical endeavor—and more. The specific environmental forces and conditions involved play important roles in such projects, and robust analysis should consider them too. Events including the settlement of land that would one day be Mexico City; its conquest and colonial exploitation; Mexican independence and subsequent nationalist and development projects; and subsequent economic booms and busts have all both responded to and informed the environmental, social, and technical conditions that spurred SASMEX creators and funders into action. I demonstrate that considering material environmental circumstances and culturally patterned understandings of earthquake risk and earthquake risk mitigation in Mexico can offer important insights for scholars and professionals in fields related to disaster response and recovery alike.

Second, I argue that there is an important contradiction in the ways that SASMEX and other technologies with similar potential benefits are often approached. While techno-optimistic accounts in both academic and popular discourse focus on how sensors or algorithmic processors will help achieve public safety, they neglect the very nontechnical elements that such outcomes require. These additional elements include, for example, considerations of how people already live with earthquakes and development of public education campaigns to help people make use of novel technologies in practical ways. Indeed, although SASMEX and systems like it are meant to change the way people live with earthquakes, the actual conditions of social life so crucial to their design and use do not get the same attention as system equipment does. Without these elements, public early warning will fail no matter how well technologies perform. I build this case as I describe how Mexican technoscience and policy have directed earthquake risk mitigation efforts, showing how the limits of their attention to and investment in SASMEX have foreclosed serious engagement with what it can really do. Through this argument, I explore the premises, opportunities, and limitations of efforts to change relationships between environment, society, and technology like SASMEX.

Finally, I argue that engineering as a field offers experts distinctive conceptual tools for environmental monitoring for risk mitigation.[7] Disciplinary logics are resources for thinking through relationships between environment, society, and technology. I show how disciplinary logics in engineering are shaped by historical and contemporary configurations of power and expertise, and how they emerge as particularly meaningful in cases of technoscientific controversy. I also demonstrate how engineering matters in the course of ordinary work of maintaining SASMEX stations in terms of technical work and the practical challenges of negotiating rural Mexico. This argument allows me to tease out some of what is distinctive about the CIRES team, and the ideas and practices they use to approach relationships between environment, society, and technology.

In the rest of this introduction, I describe how this book came to be and my goals for it. I have given a short summary of my arguments already; in what follows, I describe my research methods and the forms of data and analysis that I have used to build these insights. Next, I address the scholarly conversations that have informed this project, situating my contribution within the interests, problems, and collaborative thought of others. I articulate the strengths and limitations of this project to the best of my ability so that readers might begin with expectations of what the book can and cannot do. I close with a detailed overview of the book, suggesting how it might be read by audiences concerned with different issues.

ENVIRONMENT, SOCIETY, AND TECHNOLOGY

Earthquake early warning systems only became viable in the latter part of the twentieth century, after innovations related to automatic data processing allowed technoscientists to rapidly distinguish an earthquake from other ground motion (created by the vibrations of things like passing pedestrians or vehicles), calculate its size, and relay that information to central servers.[8] Studying such systems provides an excellent opportunity to consider how environmental monitoring technology is mobilized in risk mitigation projects.

Throughout this book, I direct readers' attention to broad categories I refer to as "environment," "society," and "technology." As an anthropologist and STS scholar, I approach these categories and the relationships among them as highly context-dependent and related to things people do, experience, and reflect on.[9] This means that the ways we navigate and make sense of environment, society, and technology are changeable; indeed, they are changing all the time. Considering these categories in the context of the Mexican earthquake early warning system's development and use allows me to both elucidate crucial issues related specifically to Mexican earthquake early warning and highlight critical concepts relevant to contemporary environmental monitoring and risk mitigation more generally.

Sociologist Sandra González-Santos writes that to provide real insights, research on technoscience and technoscientific practice must "avoid establishing a priori borders and distinctions between realms, between what can have agency . . . and what cannot, between the social and the technical, the macro and the micro, the national and the international."[10] In this book, I have refused received distinctions when possible and simply follow SASMEX, studying how the system emerged in different contexts. Data collection for this study required more than twenty months of fieldwork, primarily using participant observation, archival research, unstructured interviews, and survey research. I discuss these activities, and the work that came before and after them, in the methodological appendix at the end of this text. I also outline them briefly here.

I did fieldwork in Mexico between 2011 and 2019, which included a period between 2013 and 2014 in which I was entirely based in Mexico City and focused on this work. In various neighborhoods in Mexico City (especially Narvarte, Centro Historico, Condesa, and Xochimilco), as well as sites in the cities of Chilpancingo, Guerrero, and Oaxaca City, Oaxaca, I met with administrators, scientists, engineers, technicians, teachers, entrepreneurs, social scientists, and policymakers. I surveyed, interviewed, and learned from people interacting with SASMEX. I shared lunches, coffees, and long walks with many thoughtful people, joined them for meetings and conferences, and performed formal interviews and surveys. Access to the archives at the National Center for Prevention of Disasters (CENAPRED) and the libraries of the Center for Research and Higher Studies in Social Anthropology (CIESAS) and Iberoamerican University have been incredibly helpful. I allocated the most time to CIRES, spending eight months embedded in the offices of the nongovernmental organization responsible for SASMEX. Many of the people employed by CIRES were generous with their time and their reflections. I was particularly interested in how they worked in confusing and changeable conditions.

After I returned from fieldwork, I felt a strong sense of recognition when I read STS scholar Manuel Tironi's description of experiments that happen during disasters when "the world is uncanny and indeterminacies are excessive and radically vital."[11] Tironi demonstrates how disasters provide opportunities for novel organizing that he calls "experimental" both because they are emergent and in formation and because they involve efforts to understand the world in new ways. I found evidence of similar experimentalism in my explorations of SASMEX, in efforts to respond to past disasters, and tried to prevent new ones. Considering the essential novelty of this effort to change how people live with earthquakes helped me better understand how earthquake early warning's promises might mobilize advocates and trouble detractors.

I put these people—advocates and detractors alike, as well as those caught up in their efforts—at the center of my account of SASMEX. All of them, regardless of their relation to earthquake early warning, are involved in what disaster scholar Sandrine Revet describes as an effort to forge something shared but not homogenous—a "world of natural disasters," in which modes of legitimation, common language, and narratives are shared and debated.[12] Despite Mexico's internationally recognized leadership in disaster-related research[13] and its growing digital economy,[14] the nation lacks a reputation as a core site of innovation related to disaster risk mitigation technology or, indeed, technoscience in general. This perception has limited how widely insights developed in Mexico circulate.[15] When I put Mexico's "world of natural disasters" in center stage, I showcase insights about environmental monitoring and risk mitigation that might otherwise be overlooked by readers.

These insights are grounded in ongoing experience as well as the research methods I outlined above. In the course of my work on this topic, I have been sent out to a Mexico City street at night by an earthquake warning with my shoes still untied. I have also given talks and debated at conferences and reviewed papers. I became, in the process, a member of the expert community concerned with producing sensible, effective earthquake early warnings. While I now spend less time in Mexico than I once did, my research on this topic has never properly ended. I continue to consider earthquake risk mitigation in other contexts, and this ongoing work informs the choices I make as I write today. I collect newspaper articles, policy documents, and reports. I attend lectures and engage in formal and informal conversations with policymakers, students, entrepreneurs, scientists, and activists who have been brought together by their interest in earthquakes and earthquake safety. I also speak with and observe ordinary people who interact with earthquakes outside these expert conversations. These inform how I write this book.

SCHOLARLY CONVERSATIONS

This book is about an attempt to transform relationships between environmental, social, and technical conditions. By considering SASMEX like this, I join a growing community of anthropologists and STS scholars who explore environments, society, and technology at once. Broadly, this book is involved in the kind of project that historian Sara Pritchard has characterized as considering complex but reciprocal relationships with nature.[16] Researchers have built nuanced insights related to this topic by studying the many forms these relationships take, noting that the apparently natural conditions might be created by humans, have important effects on technologies that humans build, or both.[17]

Scholarship on how changing technologies produce knowledge about the natural world—or, as I generally refer to it in this book, the "environment"—can help us consider how we understand and live in the world. It is a basic tenet of this work that anything we might call "nature" is always already related to human ideas, practices, and experiences. This, in turn, has helped scholars consider what "nature" and "environment" mean to us;[18] specific practices of producing, applying, and circulating environmental knowledge;[19] and the politics and novel forms of advocacy and accountability that these practices may afford.[20] I draw on this perspective to consider engineering and seismicity in Mexico—specifically, in Mexico City. As historian Matthew Vitz has noted, many stories about Mexico City paint it as a city in environmental decline and a site of destruction.[21] Like Vitz, I reject that simple narrative. Certainly, the story I tell here includes destruction, but it also shows how environmental factors and forces have shaped, and continue to shape, ordinary experience in a lively city—and, to some extent, the nation and the wider world.

I use the terms "technology" and "technoscience" to describe a variety of technical practices and objects. Technoscience is an inclusive word that encompasses everything from what happens in a research laboratory to the kind of computer-based work that occurs in an ordinary office, as well as the technical labor that makes both possible. I highlight engineering as a particular kind of expertise and practice within technoscience, contributing as I do so to a lively conversation in STS about engineering identities and agencies[22] and how this form of expertise has taken its shape from, and informed, broader political and social conditions.[23] Conditions and relations involved in engineering expertise are worth interrogating, not just to better understand the development and maintenance of a single Mexican technology but to more fully grasp contemporary technoscientific work related to environmental risk mitigation.

Risk mitigation is, after all, far from straightforward. In literature explicitly related to disaster risk mitigation, risk is often a matter of mathematical probabilities regarding the likelihood and severity of potential dangers. This way of engaging with risk is useful for facilitating comparison between one potential outcome and another, or even one case and another.[24] It simplifies communication and helps decisionmakers navigating high-stakes work make challenging choices. This model of risk is, of course, historically contingent—it is just one way of talking about the likelihood of undesirable outcomes.[25] When we use the word "risk," we are calling out what we consider unusual and troublesome, contrasting it to what we consider more ordinary.[26] In this book, I am interested in risk as a way that people indicate how they see this uncertain world that we share, what they value and prioritize, and how they seek to protect it.[27] Risk is an essential part of the vocabulary that disaster risk mitigation professionals use to assess and advocate for transformations in the relationships between environment, society, and technology. While these people are, like me, members of expert technoscientific communities concerned about public well-being, I do not adopt their risk languages wholesale. Instead, I seek to contribute to a body of literature that considers their approaches to risk and disaster in practice.[28] Mexican engagement with risk is framed by environmental hazards as well as politics and the distribution of power and resources in ways that determine not only exposure but also strategies for risk mitigation.[29] With its focus on one effort to mitigate risks, the account I offer here complements other social scientific work on the always-political distribution and impact of disasters.[30]

In her book *Geontologies: A Requiem to Late Liberalism*, Elizabeth Povinelli argues that considering the everyday ways we relate to the materials that comprise the earth itself helps us understand modern configurations of power and knowledge that inform how we understand disasters and their effects.[31] My book owes a great deal to this insight as well as to other related work that encourages careful attention to how people live with the materials and forces underfoot, from metaphors[32] to industries,[33] from surface soils[34] to deep aquifers.[35] Large-scale studies of earthquakes and seismic knowledge production in particular have already drawn attention to how seismicity is a social and political force.[36] In Mexico, where earthquakes are common and violent earthquakes have had terrible effects in recent memory, seismicity is a presence in electoral politics, institutional forms, and personal practice.[37]

This book is written to participate in substantial scholarly conversations. However, I know that readers may also be interested in things far outside the scope of my work. For example, I anticipate that those involved in the practical and complicated task of risk communication may be concerned with issues related to helping share information so that people prepare for and take effective emergency actions. I can certainly recommend further reading on this topic,[38] but it is not one I address at great length in the book. While my focus is not on this issue (or, I'm sure, many others that may be important to readers), the book does advance understanding of what risk mitigation is and what it does, practically and conceptually, in the world.

READING THIS BOOK

While much of this introduction is written for anthropologists and STS scholars, I make a point to nod to people who don't fit into those categories. The truth is that I hope anthropologists and STS scholars will not be the only readers interested in this book. There are many people who are developing, advocating for, transforming, and rethinking earthquake risk mitigation technologies right now, and I have attempted to write for them, too. People invested in Mexican risk mitigation from a practical perspective may be interested in this account, if only as a record of challenging and innovative work that they and their colleagues have taken on through the years. The earthquake early warning practitioners who are building new systems across the world may pick up this book too.

The case that I make throughout this book—that environmental monitoring and risk mitigation technologies like SASMEX are efforts to transform relationships between environments, society, and technology—is specifically designed to contribute to scholarly conversations in anthropology and STS. Those readers working to develop and use earthquake early warning systems may not be deeply interested in this intellectual project. They will nonetheless find, in the following pages, a robust empirical account that includes some crucial topics rarely addressed in reports and peer-reviewed articles on earthquake early warning systems and risk mitigation work.

The book is organized topically to showcase its three major arguments. In part I, "Environment, Society, and Technology," I make the case that the development and use of this risk mitigation technology is not just a sociotechnical endeavor. In chapter 1, "Life with Earthquakes," I explore the history of earthquakes and human responses to them in Mexico from the sixteenth century to the present. Earthquake science textbooks often refer to abstract narratives about the effects of earthquakes on their environments. However, the specifics matter. Engaging seriously with ideas and practices that inform technoscientific earthquake risk mitigation requires refusing vague, imprecise narratives about what earthquakes do. Instead, I trace the specific ways that new risk mitigation tools and political action become possible in the context of hazardous environmental conditions. Chapter 2, "Earthquakes and Warnings," offers a window into what it is like to live with earthquakes as well as early warning systems. In this chapter, I consider all sorts of warning and earthquake events as part of the experience of using this kind of technology. Here I interrogate how this technology contributes meaningfully to life with earthquakes, for all that it may not utterly transform the experience of earthquakes in the way some proponents wish it would.

In "Risk Mitigation Technology," part II of the book, I explore risk mitigation and disaster risk reduction. I put SASMEX in the context of global trends and ongoing conversations to draw attention to the discrepancies between how the promises of such technologies are often discussed and the practical realities of deploying them. Chapter 3, "A Technology to Mitigate Risk," introduces earthquake early warning as both an international preoccupation and a Mexican phenomenon. I show how earthquake early warning fits into contemporary ideas about risk mitigation, exploring opportunities and challenges related to system implementation. In chapter 4, "Integrating Infrastructures," I address earthquake early warning, not in isolated terms but as part of a set of semi-integrated systems in Mexico. Doing so allows me to illustrate the dynamics of risk mitigation technology operation in practice. With a close examination of one memorable event, I explore what being responsible for earthquake early warning means.

In part III, "Engineering with Earthquakes," I focus on the experts who run SASMEX. I argue that the CIRES team draws on engineering to help them navigate their work on relationships between environment, society, and technology. In chapter 5, "Measuring Earthquakes like Engineers," I consider how the CIRES team uses engineering expertise to seek to understand and transform the ways people live with earthquakes. I describe how the team's engineering priorities play out in their technical choices regarding SASMEX. In chapter 6, "Fieldwork and New Encounters," I attend to CIRES engineers and technicians' work to maintain SASMEX. Here, I show engineering approaches to engaging with the network that early warnings rely on and the topography of the territory this network is in. I close by noting what SASMEX might do, not just for users but for those who build and maintain it. In the conclusion, I make the case for a more consistent and holistic approach to earthquake early warning. I consider the work of both scholarly and applied efforts to better understand emerging technoscientific approaches to environmental monitoring and risk mitigation. Finally, for those who are interested in better understanding how to undertake research like this, I provide an extended methodological appendix.

Contemporary discussions about these systems in public, policy-oriented, and even expert circles often focus on their technical processes and general principles for using them to prevent disasters. This book uses STS and anthropological insights to make sense of a variety of efforts to change how we live with earthquakes. Where do these efforts come from, what do they do, and what might they mean? Understanding them holistically and in context reveals the challenges that prevent such efforts from delivering on their promises.

I Environment, Society, and Technology

1 LIFE WITH EARTHQUAKES

This book must start with earthquakes and the issues that make earthquakes of high-priority interest for technoscientists, emergency managers, and Mexico City residents alike. Experiences with earthquakes, and the knowledge that more earthquakes will come, are meaningful in many different ways. One way or another, earthquakes reveal and transform organizations, regulations, and cities. This is why historian Giacomo Parrinello writes about earthquakes having a voice that "resounds in the history of places."[1] I had understood this, to some extent, but that voice had always seemed distant to me before 2017. The earthquake that shook Mexico City that September marked my first personal encounter with a big one. Its voice roared.

Though exact numbers are hard to come by, more than two hundred people were killed in Mexico City alone—and almost 150 elsewhere. An estimated forty-four buildings fell.[2] I waited a month out of respect for the upended lives and mad rush of recovery work before I booked tickets. I had hopes that what I learned could help me understand how people live with earthquakes. So I went. I found a flight to Benito Juarez Airport, open enough to welcome visitors, despite everything. I talked earthquakes with everyone who was willing, starting with the taxi driver who drove me from the airport to La Condesa, the area hit the worst.

The magnitude 7.1 event's effects were still fresh. On the ride, I learned that the taxi driver had been startled by the quake but lived nowhere near the affected zone. His family was safe. As we approached the neighborhood itself, neat piles of debris sat on street corners waiting to be cleared. Many of La Condesa's elegant art deco buildings had been closed, pending repair, with official notices related to their condition pasted across their doors. The normally bustling neighborhood was quieter than I had ever seen it.

I soon learned that some apartment buildings were wholly unlivable. Cafés, restaurants, and clubs had closed to the public, though I was happy to find that my favorite mezcalería was still providing a place for exhausted neighbors to gather and organize. After the earthquake, the tourists who would usually jockey with locals for a corner of the dark and beautiful space to eat blue corn quesadillas and drink artisanal spirits had gone elsewhere. It seemed to be mostly locals there now—and me, of course.

I had never seen the impact of such a severe earthquake before in person. Its physical evidence was everywhere—crumpled buildings, broken windows, and memorial altars of marigold flowers and candles (as shown in figure 1.1). Tree roots angled out of the ground in Parque México, the manicured green space at the center of the neighborhood, levering the

FIGURE 1.1
La Condesa neighborhood earthquake memorial, decorated with marigold flowers, notes, streamers, and candles. It is decked with the message "Fuerza México," which could be translated to read "Stay strong, Mexico." *Source*: Author (2017).

paving stones of the pedestrian walkways aside. To use Parrinello's words, I found the "voice" of the earthquake incredibly loud but still hard to parse. I thought I saw the earthquake's effects, but there were some things I was unsure about. Had the paving stones been thrust up by the most recent earthquake or an earlier one, or had they slowly become like that over years? Even when I knew that I was seeing recent earthquake damage, I had questions about it. Surely it was not the earth motion alone that cracked buildings but also the quality of building materials, design choices, deferred upkeep, or public works funding priorities. What was I really seeing when I looked around me? What kind of sense could I make of it?

My friends Beca and Enrique González illustrated the complexity of tracing earthquake effects when we strolled through the neighborhood together during my 2017 visit. The couple had lived in La Condesa for decades, and they took time off from their efforts at neighborhood recovery work to walk with me to a pastry shop across Insurgentes Avenue. As we ambled to our destination, they showed me damage and debated its interpretation. The roots of this or that tree could have been newly exposed, certainly, but it was hard to remember what the sidewalk had looked like earlier that month, when a quake originating in Oaxaca had shaken the city, and much harder to remember how it looked the month before that. Soil effects unrelated to earthquakes could have pushed them up and out of the ground. It was likely to have been a combination of things—perhaps pressure underground had pushed the roots upward for years before the 2017 earthquake finally heaved them into the open. Damaged structures prompted the same kind of discussion. It was hard for my friends to recall the precise degree to which buildings had been cracked or leaning before the earthquake. Researchers were investigating the topic, but Beca and Enrique did not refer to the meticulously collected structural assessment data that was circulating at the time.[3] They recalled their personal observations and experiences in order to situate this seismic event in physical and social context. Even in the case of a leaning house that they remembered had been standing straight before September 17, they had questions about what had actually caused the damage. Had the building crumbled because of its own seismic response, or had a neighboring structure, built in defiance of the city's codes, bashed into it? Even if this had not been true of the specific house they wondered about, their questions revealed their serious concerns about government and community responsibility for public safety.

While this may have been my first direct experience with an earthquake so severe, it was not Beca and Enrique's, and certainly not Mexico City's. Beca and Enrique's concerns were well merited in the context of Mexico City's history. Earthquakes have informed, broken down, and built up the massive city's environmental, social, and technical systems. Living with seismicity means living not just with earthquakes but also with their many effects in the physical world and in social life. The way that Beca and Enrique's speculations on the source of the damage around them involved diverse physical forces as well as government regulation and the personal choices of building owners demonstrates just how thoroughly earthquakes, even big ones, are integrated into other aspects of Mexican life. Understanding how earthquake risks inform mitigation efforts means taking these integrated effects seriously rather than attempting to artificially isolate some earthquake effects from others.

In Mexico, these histories stretch into the distant past. Mesoamerica is the "land of the moving earth" according to anthropologist Eric Wolf.[4] In *Sons of the Shaking Earth*, his sweeping 1959 overview of the histories, geographies, and cultures of the region, Wolf writes about earthquakes as a condition of life that unites diverse peoples. Stretched along the Pacific "Ring of Fire," Mesoamerica is a seismically active place.[5] In the past century alone, the territory within the national boundaries of Mexico and just off its coast has released more than ten thousand quakes measuring over magnitude 4 and forty-four quakes over magnitude 7.[6] Earthquakes endanger almost thirty-three million people in Mexico. More than a quarter of the nation's population lives in areas designated very high risk due to the severity and frequency of earthquakes in the past and the probability that this seismic activity will continue,[7] a useful consideration when discussing how earthquake risk has emerged as a national priority there (see figure 1.2).

Earthquakes have effects beyond the release of seismic energy through soils. They are evident in a variety of ways from building damage to social organization to how technologies develop. In this chapter, I show how earthquake early warning technology became possible in Mexico because of accreted effects of lava flows and colonial hydroengineering, ongoing seismicity and one particularly disastrous quake, a long social activist movement, and the political will that sprung from a fumbled response and recovery effort. Here, I interrogate how new technologies and social action become possible in the context of environmental hazards (and sometimes vice versa).

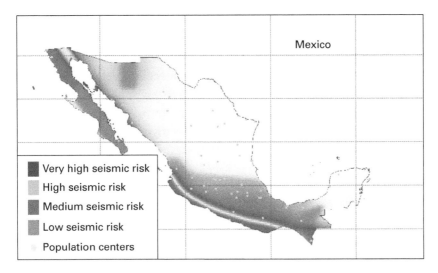

FIGURE 1.2

Map of seismic danger in Mexico, using maximum soil acceleration with a 475-year period of return. Mexico is marked by bands of shading, indicating high risk on the western shore of Mexico. Cities around the nation are picked out as points of color. Mexico City is in the low-risk area, its unusual zone of seismic sensitivity too small to be seen at this scale. *Source*: Mario Gustavo Ordaz Schroeder (2021).

I describe the city's political geology in relation to pre-Colombian settlement, Spanish conquest, and ongoing urban development. I address the physical and social effects of the disastrous earthquake of 1985 in this context, then go on to describe how the earthquake early warning system in Mexico was designed in response to deep and more recent histories. Finally, I return to 2017 to demonstrate the complex effects that these conditions still have for earthquake risk mitigation strategies and the technology developed to support it.

Seismicity has powerful effects; for that reason, stories of simple environmental determinism implying that earthquakes themselves forced technical or social change are tempting. Investigating the ideas and practices that inform technoscientific earthquake risk mitigation requires refusing such vague, imprecise narratives about what earthquakes motivate or how people live with seismic disturbance and instead considering the stories that people involved in risk mitigation tell. As scholars like Ashley Carse[8] and Sara Pritchard[9] have demonstrated, the conditions that make technological systems desirable, or even possible, are not only within the control of human beings. These conditions may be as hard to track as the complex story behind a paver's rise or a building's crack but are nonetheless both essential to, and transformed by, technical projects. Following such insights, this chapter interrogates the particular effects of historical environmental, social, and technical relationships to offer new perspectives on environmental monitoring and risk mitigation. The first step to doing so is grappling with how seismicity's complex effects unfold within Mexico's political geography.

A SENSITIVE PLACE

Mexico, situated along tectonic interfaces and seamed with faults, is subject to earthquakes. I chose to base my research on earthquake risk mitigation technologies in Mexico City because of how earthquakes have driven technological risk mitigation there. Today, Mexico City proper is home to nearly nine million people, with over twelve million more living in the surrounding metropolitan area. While the city's sensitivity to earthquakes results from unique geological conditions, social history has also framed the conditions that people live with from the ground up.[10] As a seismic city, its geology and political geography are entangled.[11]

Mexico City sprawls across a high plateau over 7,000 ft above sea level. Volcanic materials laid down over sixty-six million years encircle it—flows of lava, fields of volcanic rocks, and deposits of ash. These overlay limestone from the Cretaceous period, which dates to perhaps another hundred million years earlier. A chain of volcanoes rings the plateau. Among them are Popocatéptl and Ixtaccíhuatl, both easy to identify when they become visible through the city's smog.[12] These million-year-old volcanoes delineate a barrier between the Valley of Mexico and the Valley of Puebla. It was only fifty to one hundred thousand years ago that the Sierra Chichinautzin volcanic field closed the Valley of Mexico off and created the conditions for a basin that would become marshy ground.

The five lakes that came to occupy that ground are a crucial part of conversations about seismicity in Mexico City. Lakes Zumpango, Xaltocan, Xochimilco, Chalco, and Texcoco are no longer present in the city the way they once were, but they still inform how the ground responds to seismic motion and how technoscientists talk about the city as a seismic place.[13] Mexico City was built on top of the Mexica city of Tenochtitlan, a center of power in what is often called the Aztec Empire.[14] The Mexica built on water, creating platforms and networks of dams and causeways (see figure 1.3). This infrastructure, arranged radially, connected Tenochtitlan and its twin city Tlatelolco to settlements throughout the area. Secondary causeways and aqueducts regulated water flow and travel.[15]

These elaborate hydroengineering projects were already under way in the fourteenth century when nomadic Mexica people arrived in the valley and saw, if legend is true, the eagle eating a snake that is now emblazoned on the Mexican flag. One hundred years earlier, Xaltocanmeca communities had begun building artificial horticultural beds raised above the surface of the lakes called *chinampas* and leaching salts from brackish lakes to improve growing conditions. As the Mexica's Triple Alliance united the twin cities of Tenochtitlan-Tlatelolco with those of Tlacopan and Texcoco to rule what some estimate to have been ten million people in Central Mexico, they expanded these preexisting water projects. By the time a coalition of Spanish forces and rival Mesoamerican states conquered the city in 1521, Tenochtitlan was already two hundred years old and had achieved hegemony within the Triple Alliance. The invading Europeans called the city the "Venice of the New World."[16]

The city's political status and the sophisticated hydroengineering of these waterways made the Spaniards' decision to set up their capital on top of Tenochtitlan advantageous. In her analysis of the ways class informed and was enacted through the conquest, Vera Candiani attributes their choice to both the city's luxuries and the strategic barriers to attack that its waters provided. It was not long, however, before the conquistadores became frustrated by the periodic flooding of the same hydraulic engineering that had at first astounded them and undertook an ambitious effort to drain the lakes.[17] Although flood control was an original rationale for the drainage project, one of the Spaniards' first attempts to drain the waters instead flooded Tenochtitlan for five years. Efforts to keep the area dry have been in motion more or less constantly since 1607.

Seismic, social, and technological conditions informed each other as the lake water was channeled away and the city grew. The drainage project allowed Spaniards to eliminate the Mexica infrastructure and expand their own colonial city, but it had unfortunate consequences for the city they developed. By draining the lakes, the Spanish exposed the fine floating volcanic ash and silt that had accumulated on the lake beds. Under the

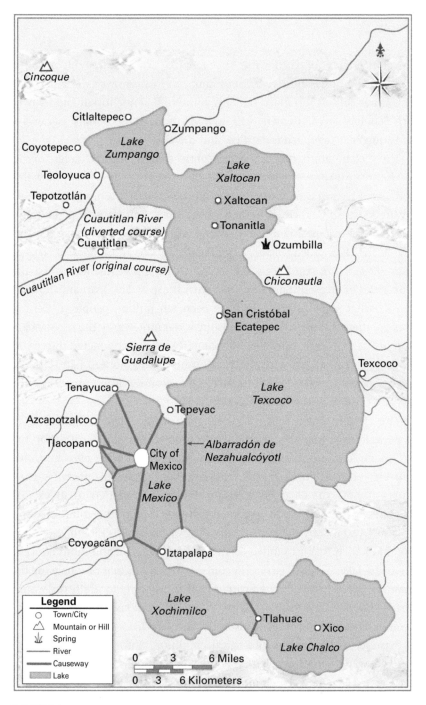

FIGURE 1.3

The basin of Mexico and its major hydraulic structures in the sixteenth century. Originally published in Candiani, Vera S. *Dreaming of Dry Land: Environmental Transformation in Colonial Mexico City.* ©2014 by the Board of Trustees of the Leland Stanford Jr. University. All rights reserved. Used by permission of Stanford University Press, sup.org.

right circumstances, these soils are subject to an effect called liquefaction, in which they take on the properties of liquid under certain circumstances. When soils like these are saturated with water and vibrate at certain frequencies, they may lose their capacity to bear weight and behave more like water than like solid ground. The structures built on them can sink unevenly into the earth. But these soils do more than that. Earthquakes propagate faster through hard rock than sediment. However, muds, sands, and silts (particularly when water-saturated) not only make seismic waves bigger but also extend their duration. In other words, the very kinds of soil that Spanish drainage exposed in Mexico City amplify the strongest earthquakes.

The implications of these conditions are fascinating and dangerous. Mexico City is not close to many seismically active areas, but the soils beneath it are among the most sensitive types documented by geoscientists. Mexico City shakes, and then its shaking increases vibration in surrounding material. This effect is severe: Mexico City sometimes registers seventy-five times greater motion than places on hard rock located an equal distance from an earthquake's epicenter.[18] Hundreds of years after the Spanish conquest, a serious geophysicist at a research institute used a graphic modern metaphor to communicate about how it did so. The city is, he told me, like a shot glass of Jell-O on a shaking desk.

Civil engineers and geophysicists document the seismic amplifications caused by that soil.[19] In maps like the one in figure 1.4, areas that are especially sensitive to earthquakes are part of the "lake zone." These are distinguished from the hard-rock volcanic areas designated part of the "hill zone," which is less responsive to earthquakes. Between them is the "transition zone," with soil that mixes fine particulate and volcanic rock, silt, sand, and clay.

The Valley of Mexico already had a substantial record of dangerous seismicity before Spanish settlers drained the lakes and built the city up. Indigenous Mesoamerican depictions of seismicity are famous for symbolically associating earthquakes with cataclysmic changes in human lifeworlds.[20] Nahuatl-speaking people like the Mexica, for example, tracked the earthquakes they experienced as well as those they expected in the future using a combination of glyphs for *tlatli* (earth) and *ollin* (movement). Rendered with different colors and configurations, these depictions recorded seismic action, showing information related to the time of day and potentially even

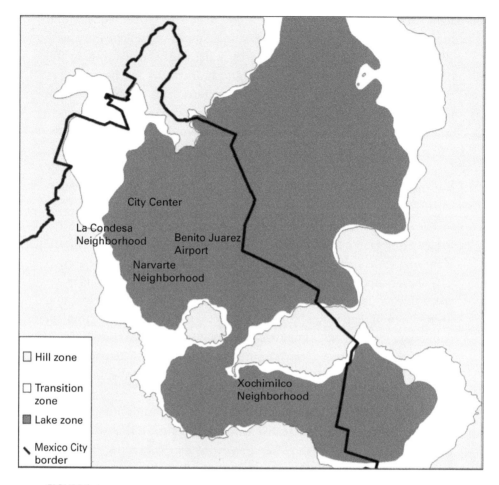

FIGURE 1.4

Mexico City soil zones, with areas of particular interest for this book labeled.
Source: Mario Gustavo Ordaz Schroeder (2021).

the intensity of individual quakes (see figure 1.5).[21] From these documents, we know that the earth shook people in the Valley of Mexico often and sometimes disastrously. There is evidence, for example, that a violent earthquake damaged many buildings in Mexica-ruled Tenochtitlan in 1475.[22]

Spanish records of life in Mexico City also suggest that the colonists frequently experienced disturbing and sometimes deadly quakes.[23] An earthquake in 1711 was reportedly so severe and extended that it took nearly half an hour for the ground to stop moving.[24] Others, like the ones

FIGURE 1.5
This nahuatl glyph *tlalollin* from the Codex Telleriano-Remensis indicates a nighttime earthquake that analysts suggest took place in the year 1480. *Source*: Espinosa Aranda et al. (2009).[25]

recorded in 1768[26] and 1800,[27] caused severe damage to buildings and even deaths. These experiences came to inform architectural choices as the city developed, but the new styles of building inevitably led to novel, and sometimes unpredictable, experiences with shaking.[28] By 1970, there were six million people living in the city; by 1980, the population had grown to eight million.[29] Mexico City's transformations created new opportunities for seismic danger and new ways that it could take on political meaning.[30]

1985

Mexico City has remained both sensitive to earthquakes and nationally important throughout its accumulated seismic and social history. In 1985, Mexico City produced 44 percent of the Mexican GDP and was home to one-fifth of the country's total population and one-third of its public employees.[31] It housed the nation's premier universities, museums, and governing institutions. Its history of entangled physical and political upheavals further developed that year when a series of earthquakes ripped out of the subduction zone off the coast of Michoacán and shook the city that September, disrupting the economic, emotional, and administrative lives of people across the nation. Journalist Jacobo Zabludovsky compared the event to the violence of Spanish conquest and settlement. He underlined resonances between traumas brought about by threatening environments and by human beings themselves, describing the quake as the largest tragedy in the history of the capital, "preceded only by the destruction of the Aztec city at the hands of the conquistadores in 1521."[32]

The first earthquake happened on September 19, at 7:19 in the morning, measuring magnitude 8.1 at its source. In the lake zone of Mexico City, almost 230 miles to the east, the ground moved back and forth a full eighteen inches every two seconds for three full minutes. The second quake happened the next evening, at 7:38 p.m., measuring at magnitude 7.9.[33] The seismic energy that the two earthquakes released caused landslides that damaged buildings and infrastructure across Mexico.[34] The way the earthquake unfolded was unprecedented, to the knowledge of many scientists. When Mete Sozen, a civil engineer at the University of Illinois, testified before the US Senate on the disaster, he said:

The amount of energy that this earthquake pumped into their high-rise construc-
tions was, I think, at the extreme edges of the incredible region, and I think that
this is probably the main lesson I have learned from this earthquake, that I find it
easier to believe the extremes of credible events now than I did before.[35]

Just as no other place experienced these levels of destruction, however,
the earthquakes did not inspire political and technoscientific work elsewhere
as it did in Mexico City. Neighborhoods in the lake zone of Mexico City
were hit quite badly, including the city's historic downtown, the residential
neighborhoods of Roma and La Condesa to the west, Tlalpan and Xochimilco to the south, and Tlatelolco to the north. All in all, the earthquake
damaged around three thousand buildings. More than 750 fell. Among
these were hospitals, large apartment complexes, hotels, and key nodes of
administrative infrastructure. At least six thousand people lost their lives
in Mexico City alone, although estimates vary drastically and disturbingly;
the true number may be many times that.[36] Not only were people hurt, but
the means to treat their injuries were limited, as were the organizational
capacities of city and state authorities. One estimate suggests that 120 government agencies lost parts of their facilities.[37]

Nearly thirty years had passed by the time I first arrived in Mexico and
began asking questions about life with earthquakes and how it might
inspire technologies. But for all that distance, the 1985 event still came
up often in my interviews with older Mexican technoscientists and emergency management professionals. Elisa Alonso, a semi-retired risk mitigation professional who had made a career advocating for earthquake safety
in government agencies, had not suffered harm herself, nor had anyone she
was close to, but the chaos of postquake infrastructure breakdown gave her
family cause to worry. She told me:

> My father and mother were in Spain, and the news they heard was that Mexico
> City had been destroyed . . . and they didn't know. All of us, all of their children
> were here in Mexico. It was horrible. It wasn't as easy to communicate then as it
> is now. The telephone network . . . had collapsed like a house of cards. It left the
> city incommunicado. The conclusion was that everyone was dead. And indeed,
> many were.[38]

The Mexican state emergency response received heavy criticism and still
troubled most of those I spoke to about the event.[39] Officials were absent,
or at least appeared to be. It was not until thirty-nine hours after the first

quake that President de la Madrid finally addressed the Mexican people on television, and his reputation suffered for it.[40] At the time the earthquakes struck, Mexico did not have a dedicated disaster response organization or even a substantial disaster plan for the capital city. Instead, the federal government had something in place called the DN-3, which simply assigned the Mexican Army responsibility for coordinating action in case of such a disaster.[41] But because residents and politicians alike were wary of empowering the army, they could not respond effectively to the unfolding crisis.[42] Army deployment remained limited, and volunteers resisted working with them. Soldiers were restricted to providing security and crowd control in Mexico City.[43] Since help from the state was lacking, ordinary people organized themselves to respond to the city's emergency.[44]

In 1985, many Mexicans had experience with activism. The country was already in the middle of an economic crisis that affected ordinary people in their daily lives.[45] By 1982, only three years before the earthquake, Mexico was carrying $97 billion USD in foreign debt. Unable to pay its debt service, the country agreed to significant structural and free market reforms including a reduction in public services. In response, Mexicans, and particularly residents of Mexico City, became increasingly politically active as IMF-imposed austerity measures started to kick in. City residents had mobilized to demand the housing, transportation, and other services that they had come to expect.[46] After the quake, Mexico City residents organized themselves at an entirely different scale, undertaking search and rescue, shelter and care, and other emergency tasks locally and in ad hoc arrangements. Two million people took part in response and recovery work, one out of every eight adults living in the city.[47]

Many people in Mexico City had experience with organizing. However, when I asked people who lived through this time about how such experience served them, they cautioned me not to make too direct a comparison between what people had been doing to oppose austerity measures before 1985 and what they did in the wake of the earthquake. The latter was simply *more*: more emotionally involved, more all-consuming, more transformational for their ideas about the world around them, their relationships with their neighbors, and their political lives. The experience of living through the earthquake and recovery was often challenging to narrate. It had been so different from ordinary experience. Elisa Alonso, the same former risk mitigation professional who narrated her Spanish parents' panic earlier in this chapter, summed up her own experiences with a single telling sentence: "People were crying, and we held each other," she said. She continued: "It was simply emotional. We had to control our own fears. It was horrible, what we saw of that earthquake. Just *horrible*, terrifying."[48]

People and donations flooded in from around the country.[49] The labor was, from all reports, intense and consuming. "Our volunteers, in the first days especially, did not rest and worked ceaselessly day and night," reported Emilio Díaz Cervantes, a founding member of the Topos emergency response team, in a reflection published after the fact.[50] Many accounts describe a kind of transformation. Mexicans did not see their government taking action. They did, however, note the kind of organization that their own communities were capable of. In loss and systemic failure, many saw a revelation, and evidence of new possibilities. This way of understanding the earthquake and what happened next animates the compiled testimonies of earthquake survivors found in archives and bookstores today[51] and continues into ordinary conversations about the earthquake and its effects. Mexican journalist and critic Carlos Monsiváis explained how this was possible:

> The fear, the sheer terror of what happened to our loved ones and property; the loss of our family and friends, the rumors, the misinformation and feelings of impotence, all of it—changed mindsets. All of a sudden, a new idea became possible: a real civil society. And that is where solidarity starts.[52]

Many suggest that the earthquake did serious political work. After 1985, previously unimaginable political scenarios—such as taking a corrupt government to task or even ousting the Institutional Revolutionary Party (PRI), which had held the presidency since 1929—became properly imaginable.[53] In 1988, the left-wing presidential candidate Cuauhtémoc Cárdenas nearly won the election. But while PRI politician Carlos Salinas de Gortari defeated him, many opposition candidates won their legislative elections.[54] The PRI's hold on national government had become a good deal less certain. Other changes came more slowly. In 1997, a dozen years after the quake, leaders of Mexico City began to be elected by citizens rather than appointed, giving those who had felt neglected in emergency response more direct say in city governance.[55] Three years later, the Mexican people unseated PRI from the presidency.[56] Despite the passage of time, these events are often associated with the disaster in 1985 and the solidarity that Mexico City residents found among their communities and their frustration with political leaders.

The earthquake also had effects on Mexican social life. In the wake of the earthquake, however, Mexico City cartoonist Alberto Beltrán wisely told reporter Elena Poniatowska, "Humans do not change only because an earthquake has shaken the earth. Mexican society moves slow, little by little, not in leaps." Beltrán's reflection provides a good tool to think through seismicity's effects. Environmental hazards do not have effects outside of particular physical and social historical contexts. They can, however, inform new efforts in risk mitigation. The resources and interest that the 1985 earthquake catalyzed had consequences—particularly for the technological project on the south side of Mexico City that would become an earthquake early warning system. Situating technological development in the context of environmental and social conditions in post-1985 Mexico City allows me to demonstrate the contingent nature of this system's form and relate its growth to specific issues in the city and nation.

ENVIRONMENT, SOCIETY, AND TECHNOLOGY AFTER 1985

Even though the Mexican state may have fumbled its immediate response to the 1985 earthquakes, it was able to marshal resources and political will for recovery efforts. In the aftermath, there were significant national and international resources available for projects that drew on scientific expertise to better understand what had happened and prevent a similar disaster from taking place again. Some estimate a reconstruction cost of 6 percent of the nation's gross national product at the time, and there was significant space in that budget for new risk mitigation projects.[57]

Reflecting on this period, the Mexican experts I interviewed describe a time of heady excitement, as projects that had been in the works for years received an influx of attention and funding. One emergency management consultant was still able to describe a colleague's enthusiasm as Mexico began to pay serious attention to seismic risk: "The prophecy of his life had been fulfilled. He would say, 'I'm embarrassed, people think I'm enjoying myself. I'm not enjoying myself. But I've studied this for years and it is thrilling.'"[58] Mexico City was not only a site of severe seismic vulnerability and destruction but also home to researchers and technoscientists who had built their careers studying these things.

In October of 1985, the Mexican president Miguel de La Madrid convened a National Reconstruction Commission. Along with taking on the monumental tasks of addressing disaster response and recovery, the commission was tasked with assessing risks and developing policy recommendations for nationwide reform.[59] These efforts were quickly funded, manifesting newly focused attention on disaster prevention and risk mitigation and creating new annual emergency recovery funds. While Mexican national, state, and local governments made space for a National Civil Protection System, national and international institutions offered funds for research on earthquakes. Things were changing quickly, and there were resources available to support new ideas.

Out of this energy, new technologies and organizations emerged, exploring possibilities for different ways to live with earthquakes. Some of these efforts would result in Mexico's public earthquake early warning system and the NGO responsible for it, Center for Seismic Instrumentation and Registry (or CIRES, from its Spanish-language initials). Engineer Juan Manuel Espinosa Aranda would be at the center of both from their inception to the time of this writing. Espinosa Aranda was one of the many people caught up in this wave of support for earthquake recovery and disaster prevention. He was already working at the National Autonomous University of Mexico, or UNAM, in 1985. He had studied seismic instrumentation, earning his MSc working with other electrical engineers in the basement of the Engineering Institute and developing tools to register, transmit, and record data about earthquake motion in real time under Dr. Humberto Rodriguez.

The telemetric system that Espinosa Aranda worked on in Rodriguez's lab allowed sensors throughout the region to record real-time data on a reel-to-reel audio recorder and digitize it. In the wake of the 1985 earthquake, Espinosa Aranda and his colleagues took on the new responsibilities of building and maintaining seismic instruments. Their experience in Rodriguez's lab gave them the expertise to support the many technoscientific research and disaster prevention efforts that developed after the quake. Encounters with the unique telemetric system in Rodriguez's lab also inspired the earthquake early warning project that Espinosa Aranda and his colleagues would eventually develop and champion.

Even before they began working on earthquake early warning, these engineers were in demand. They had become experts in maintaining seismic instrumentation in Rodriguez's lab, so Mexico City tapped Espinosa Aranda and his colleagues when it funded a new citywide seismic data collection initiative. This project was only one aspect of post-earthquake efforts to better understand the city's seismic sensitivity and prevent similar disasters in the future. In this context, Espinosa Aranda's team was tasked with installing a network of forty accelerometric instruments within Mexico City. When they measured seismic effects around the area, researchers and engineers could better understand the city's curious geological disposition and fine-tune their maps of microzones of the geology that varied significantly throughout the area. The work had direct implications for new building regulations. Espinosa Aranda and his team formed CIRES in 1986 as an independent nonprofit entity, funded by the National Reconstruction Fund and administered by Mexico's National Council for Science and Technology (CONACYT), and began to work on risk management technologies that produced knowledge about earthquakes.[60]

At first, the newly constituted CIRES team focused only on this work, specifically the Mexico City Accelerographic Network, which was up and running by 1987 and still collects detailed data about ground motion and makes it publicly available. Soon the goals of CIRES broadened to include other projects. Using their own time, resources, and experience with networked sensory devices, Espinosa Aranda and the CIRES team developed the first version of an earthquake early warning system for Mexico City once they had the accelerographic network operating.

CIRES took direction from a report published in the immediate wake of the 1985 quake.[61] There, geophysicists suggested that another large earthquake would likely shake Mexico City again soon. Working on the theory that pressure in a fault would build up until it was released all at once in a large seismic event, these scientists had identified a stretch along the Costa Grande of Guerrero that was overdue for a dangerous earthquake.[62] CIRES sited twelve seismic stations near the designated area, and Espinosa Aranda developed the system's first algorithm—what now seems a slow ten-second process—to analyze the first moments of earth motion emerging from this area. His process forecast the likely size of an earthquake based on those first moments, distinguishing between quakes likely to be small, moderate, or large. It then automatically sent the results by radio to central servers in Mexico City.[63] If all went well, the signal could arrive roughly sixty seconds before a significant earthquake, and could trigger broadcast warnings if necessary (see figure 1.6).

Work on this new technology meant that CIRES had a new project and a potentially novel way of approaching earthquakes. They were no longer

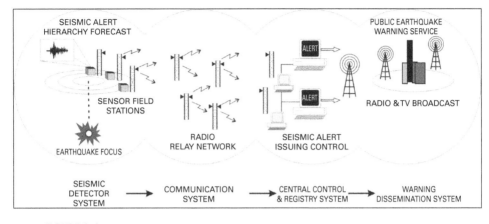

FIGURE 1.6
SAS and SASMEX function diagram. *Source*: Espinosa Aranda et al. (2009).

just collecting data for scientific research or to inform structural safety regulations. They were making a system of instruments to exploit the speed of telemetric connections to give people time to take shelter before an earthquake struck. Now they had a chance to make unpredictable earthquakes more like tornados or floods: that is, more like other hazards that people had some warning about. It was a small change, but an important one, and CIRES engineers presented the system to the Secretaría de Obras of the Departamento del Distrito Federal (DDF) in December 1989.

The engineers' aspirations for this new kind of public management work were contingent on the needs and goals of other people and institutions. They had substantial support from the charismatic Mexico City head of government Camacho Solís, but that presented challenges as well as opportunities. A longtime administrator and leader in Mexico's seismic risk mitigation community described how the earthquake warning system was promoted by what seemed to be a political whim: "[Reporters] were interviewing the governor[64] of Mexico City," he explained. "And he said that they had been working with CIRES on this alert system and they said, hey, and why don't you do this publicly? And then it was decided, in the news, to make it public."[65] Just like that, a key decision had been made. This opportunity to advance risk mitigation technology had appeared suddenly, out of the blue, and did not allow for slow and deliberate action or careful outreach campaigns.

In 1990, the Sistema de Alerta Sísmica, or the SAS, had twelve field sta-tions, and their signals were used to automatically slow subway trains when they detected an earthquake.[66] In 1991, it broadcast its first public warning out to anyone in Mexico City with radios tuned to the correct frequen-cies to receive it. But as promising as the system seemed to its proponents, direct automatic applications within technical systems were often imprac-tical.[67] It might be useful to format flexible human responses, but—with the exception of the city's subway system—it was not easily integrated into automated infrastructure. Instead, SAS was understood to produce an advantage for human action.

The city's official earthquake early warning system soon began to share alerts through partnerships with local television and radio stations and by way of technologies that CIRES engineers developed. The CIRES team designed, built, and installed SASPERs—Personalized Earthquake Warning Systems—in buildings so that sirens could go off automatically whenever SAS sent out an alert broadcast. Over the decade that followed, SAS slowly developed the capability to reach more users in Mexico City. In 1993, the system began sending alerts to radio and television stations for broadcast to the whole city. News reports from the time show how it sounded out before earthquakes coming from the coast, giving anyone who heard the alert time to take shelter.[68] One study even suggested additional utility: when schools used the warnings, allowing children to prepare themselves and take protective action, it reduced their anxiety.[69]

Just as Mexicans' experience with earthquakes could facilitate support for developing the earthquake early warning system, those same experi-ences could also undermine public trust in the system when funding for the project flagged. With just twelve stations strung along the Pacific Coast, the SAS could not detect or alert for all the earthquakes Mexico City expe-rienced. Plenty of quakes came with alerts, but others arrived in the city without any announcement because they originated far from the small sensor array or fell under the threshold deemed large enough to alert.[70] Institutional support for CIRES was deeply uncertain. The government had invested an estimated $600,000 USD for design and development, but then its support for the system—both for maintenance of technical com-ponents and the work necessary to help Mexicans understand how to use the alert—dwindled away.[71] Mexico's Civil Protection System coordinated some educational outreach in the mid-1990s, but this was unsubstantial and short-lived.[72] Efforts to explain the system and manage user expectations, much less expand the network's reach, grew few and far between. In 1998, Espinosa Aranda had to cover the nonprofit's payroll with his own money when the Mexico City government failed to pay necessary costs for upkeep of the SAS and the accelerographic network.

CIRES found other partners to support their work when resources in Mexico City were unreliable. The NGO built another system of thirty-seven stations in the southern mountains for quake-prone Oaxaca City in 1999; in 2005 this system was united with Mexico City's SAS. Together, these comprised SASMEX, a nationwide system, with a combined forty-nine stations. The network continued to grow, enabling alerts in other cities in Central Mexico. The tally eventually included ninety-seven stations that could issue alerts to Acapulco, Chilpancingo, Morelia, and Puebla, in addition to Mexico City and Oaxaca (see figure 1.7).[73]

The system received more attention and public support around the twenty-fifth anniversary of the 1985 quake, when a series of earthquakes claimed lives around the globe. As violent earthquakes and tsunamis racked Chile, Haiti, and Japan, they resonated with Mexican memories and concerns about safety.[74] The system had been stalled, but it grew again, not just incorporating new stations and user communities but new users as well. More Mexicans began acquiring tools to use the SAS signal. Civil Protection issued new regulations requiring businesses of a certain capacity

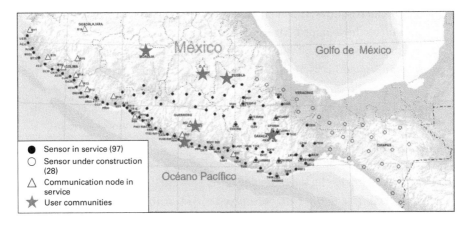

FIGURE 1.7
SASMEX stations and user communities. *Source*: CIRES (2015).

in high-risk zones in and near Mexico City to install some form of radio receiver to dispense alerts.[75] With the widespread availability of mobile phones, smartphone apps began to turn earthquake early warning messages into push notifications for subscribers' telephones and computers, setting off sirens for them.[76] Others experimented with personal alert radios that were cheaper than those previously on the market.[77] Mexico City's Civil Protection took up system responsibilities in September 2015, connecting public loudspeakers mounted on telephone poles around the city to the earthquake early warning system. By the time the 2017 earthquake happened, the network had developed substantially, still marked by the physical and social conditions of Mexico City's histories.

2017

Mexico City's particular histories and its encounters with earthquakes demonstrate just how entangled environmental, social, and technical conditions are in this sensitive, symbolic, busy city. The benefits the region offered for Mexica and other pre-Hispanic people was, by many accounts, a matter of environmental and social conditions. The Spanish conquest and subsequent occupation of the city were deeply politically traumatic and have had indelible effects for the modern Mexican nation—including radical environmental transformations as lakes were drained. These, in turn, made the city's growth onto incredibly sensitive terrain possible, and created some crucial conditions for the damage the city incurred in 1985 and 2017.

Seismicity has shaped Mexican earthquake risk management institutions and technologies, forming both the impetus to develop earthquake early warning and reasons for people to be frustrated by the technology's limitations. The seismic, social, and technical histories and contemporary life in Mexico are all closely related, mutually informing. Risk mitigation technology like SASMEX must be understood in this context because it has been supported and undermined by earthquakes and the social life of Mexico City, as should the system's mixed success on September 19, 2017, thirty-two years to the day after the 1985 earthquake that jumpstarted its development (see figure 1.8). This mixed early warning success formed part of the experiences that I learned about when I visited Mexico City and witnessed a community quieted, still mourning, and perhaps beginning to recover after another deadly earthquake.

FIGURE 1.8

Map of September 19, 2017 earthquake intensity. *Source*: US Geological Survey, National Earthquake Information Center (2017).

The city's earthquake early warning system worked, people told me, but not as well as it might have. It was the kind of complicated success that happens when a technology—designed to support public safety in light of unpredictable environmental hazards and integrated imperfectly with other systems—is put to practical use. When I told people in Mexico what I was studying, responses were varied. Many did not know about it, especially before the earthquake that shook Mexico City in 2017 launched it into popular awareness.[78] Those that did had a variety of responses. Some showed irritation about imperfect alerts, some took it in stride, and others suggested that even a partial success was better than having no alert system at all.

The complications with the warning system in September 2017 can be attributed to the relationships among earthquakes, Mexican social practice, and technologies in three different ways. The first has to do with geography and how the system was designed to prevent a repetition of 1985. CIRES engineers planned the earthquake early warning system based on the likelihood that another large earthquake would originate from the west coast of Mexico, but they did not design it for one originating near Mexico City, as this one did. It is not clear that a substantial warning could have been generated even if stations had been positioned differently, but as it was, the system did not offer sufficient advance notice before the ground began to shake.[79]

The second factor was a coincidence of timing and how the disastrous quake of 1985 is both memorialized and used as a lesson. On the morning of September 19, 2017, there was massive earthquake drill, an activity coordinated each year in remembrance of the 1985 quake.[80] The drill that morning raised popular awareness about its uses, with an estimated 7.5 million people participating.[81] Together, they responded to an "alert" at 11:00 a.m., taking action as if a magnitude 8.0 quake was rushing inland from Mexico's west coast. Some practiced evacuations. Others, particularly those working in tall buildings, moved for cover. All this activity was planned and rigorously documented for Civil Protection. Mandated compliance with these requirements was one way that national policy could encourage protective action.

The earthquake that followed barely two hours later was not part of the plan. The magnitude 7.2 quake originated outside the Central Mexican city of Puebla, less than one hundred miles away from Mexico City, not from the coast. Only a few of SASMEX's networked stations were positioned nearby. The earthquake early warning system could only generate an alert twelve seconds before serious shaking began in Mexico City at 1:15 p.m.[82] By that time, some of the fastest-moving seismic waves—the comparatively weak compression waves or P-waves—had already hit the city.

The third major reason for the mixed success of the 2017 alert has to do less with particular historical events than with ongoing choices in the techno-scientific management of life with earthquakes and earthquake early warnings. Like so many engineering projects, Mexico has approached early warning as a technical system with social implications rather than a system integrated into social life with environmental conditions. This is why some sirens blared the warning out for all to hear that afternoon, while others remained silent—their upkeep has simply been inconsistent.[83] Responsibility for producing and disseminating warnings has been distributed between agencies, with CIRES in charge of the technologies that generated the alert, but not the majority of the means by which alerts could reach people, nor public education programs to help Mexicans understand what they might hear in the second before an earthquake. These other responsibilities rested primarily with Mexico City and borough governments. This decentralized way of managing earthquake risk mitigation instead of treating it as one intervention runs counter to the experience of ordinary people. I will discuss how life with earthquakes and early warning systems unfolds in the next chapter.

2 EARTHQUAKES AND WARNINGS

My visit to Mexico City in the wake of 2017 might have been my first experience with a community in recovery after a serious quake, but my introduction to earthquake early warning systems had come years before. When loudspeakers issued an earthquake warning at 11:44 p.m. on September 29, 2015, I was on an inflatable mattress in Beca and Enrique González's home writing emails. I had two of their deep red embroidered throw pillows propping up my head so that I could make out the screen without glasses. Their living room, loaned to me for the duration of a short visit, was lit entirely by the streetlights out on Plaza Popocatepetl and my laptop monitor's glow.

It was quiet. And then it was not quiet at all. A loud digitized voice repeated the words ¡alerta sísmica!, and a siren I had only heard before in demonstrations warbled from somewhere outside. Enrique called to me from the bedroom down the hall: "Was there a scheduled drill?" He was yelling to be heard. "No, it couldn't be," I replied at volume, grabbing a coat and the smartphone charging beside it. The laptop was still on the mattress, and I would leave that, too. Where was my wallet? I found two left shoes, a heel and a flat, and searched out the match for the flat one. These had laces. Laces had to be loosened for the shoes to go on. I put my feet in them but did not tie them, and thought again about my laptop, the last time I had backed it up, and if I could afford to lose all the work saved on it if the building came down. I could not quite recall which pocket of my wheeled luggage I had put my passport in, and I was unsure if I should commit to searching all of them. Maybe I had put it in the backpack with the week's dirty laundry. The siren continued.

At the door, I met Enrique, who was fully dressed and ready to go. A moment later, Beca appeared in a bathrobe and slippers, holding their two small dogs in her arms. I was wondering about the wisdom of going back to see if I could grab my wallet when the siren stopped.

The quiet was jarring. We waited there, in the landing of their building's wide marble stairs. I counted seconds in my head. I knew that the siren should have continued to sound until the quake was over, but the loudspeakers were newly integrated into the public earthquake early warning system, and it was possible that they would not follow the same rules as the radio broadcasts I was more familiar with. Regardless, the early warning could give us, at the absolute most, a bit over a minute's advantage before an earthquake could travel from the most distant of SASMEX's sensory field stations all the way to Plaza Popocatepetl in the center of Mexico City. We were approaching that outside time limit if we had not already passed it, slow as we were to assemble at the door, but I had felt nothing.

There was indeed an earthquake on the evening of the 29th. According to the report that the Servicio Sismológico Nacional issued the next day, an event of magnitude 4.6 originated about forty-nine kilometers northeast of the city of Ometepec, Guerrero (see figure 2.1). Though it was not quite large enough to shake the second floor of Enrique and Beca's building over the La Condesa neighborhood's sensitive soil, it still triggered SASMEX. The quake was real, but the alert was inaccurate.[1] A different sort of automated analysis might have distinguished this quake from one that would actually be felt throughout Mexico City. The techniques used by the engineers running SASMEX, however, prioritized speed over accuracy, so they sometimes produced warnings about this sort of earthquake. The engineers at CIRES had explained publicly and repeatedly that these sorts of things simply happen. They had made choices for SASMEX that reflected their priorities. These priorities did not include devoting time for very nuanced assessments of earthquake magnitude or developing lengthy alert communication about the size an earthquake was likely to be or how long listeners might have before it hit them. Because of these choices, when the sirens sounded into the night on Plaza Popocatépetl, we could not know what sort of experience to anticipate. The earthquake might have been a little bigger, and we might have felt it. If we had, it might have been begun shaking us as we stood at the top of the building's stairs or even as we collected our things and prepared to leave the apartment.

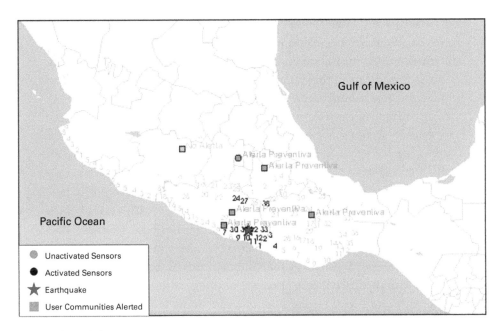

FIGURE 2.1
Map of September 29, 2015 earthquake detection and warnings. Mexico City is one of the places marked with a square, and the epicenter of the earthquake is marked with a star. Activated stations that detected motion are indicated by black numbers, while unactivated stations that did not are pale gray. *Source*: CIRES (2015)

When people live with both earthquakes and warning systems, they live with many possible configurations of the two. They experience alerts for earthquakes they feel and warnings for those they cannot without knowing for sure which any given howl of *¡alerta sísmica!* might foretell. Although emergency management professionals have shown quite convincingly that warnings for events that never happen can be considered drills or opportunities for training,[2] and that post-alert messaging can support public confidence in a system even after trouble,[3] there is a substantial difference between such a suggestion, in theory, and how people experience and respond to warnings and earthquakes in practice.

In the days that followed the late-night alarm in 2015, the siren that sounded from the loudspeakers was a topic of significant interest among people I spoke to in Mexico City. It was covered in news reports, and CIRES and other organizations associated with the system were tagged in diverse social media responses. Some people accepted this kind of warning-without-event as the price they paid for an earthquake early warning system, but many expressed frustration and confusion about how the system worked, as well as worry about the real physical harm that the shock of this warning might cause or habituation that might result from false alarms.[4]

An earthquake early warning system is a tool built to intervene in users' experiences in just one moment, to allow certain opportunities for diminishing risk.[5] In this moment, we can have dogs and shoes ready to go, laptops and wallets either at hand or left behind. Such technologies are easy to break down into components or subsystems, making a unified whole into separate processes with different responsible parties. This is very much how Mexico's earthquake early warning system has been treated. CIRES was responsible for seismic instrumentation and registry, for generating alerts to the specifications of the cities and private businesses who contracted with them. Civil Protection was in charge of distributing alerts, and local authorities were responsible for maintaining loudspeakers. General education and outreach was largely theirs, too. However, the integrated experiences of people who live with earthquakes and early warning on a day-to-day basis are harder to partition. We respond to our experiences, our knowledge, and encounter every new warning as a single intervention: a siren in the night rather than a variety of subsystems performing to different standards.

In this chapter, I put my early warning experience in the context of life with both earthquakes and warnings, treating earthquake early warning as a heterogenous system.[6] I choose to describe this case precisely because it is so far from ideal but nonetheless absolutely ordinary. Starting from here allows me to show how people live with earthquakes and technologies in ways that include earth-shaking of many different intensities, emergency education, and, of course, the particular technical tools and decisions that comprise SASMEX.[7] For all that many might consider this an imperfect orchestration of environment, technology, and social action, alerts like this are just as crucial to risk mitigation as cases in which a sizeable earthquake triggers a warning. Here, I interrogate Mexico City's current normal: ordinary experiences of life with ongoing earthquakes and various kinds of earthquake early warnings. It is not as different as the most optimistic of early warning advocates might wish. However, it does entail particular tangles of collaboration and confusion. Considering these can help us better understand the kinds of collaborations required to grapple with and change the way people live with earthquakes in Mexico.

I first consider what it means to address earthquake early warnings and earthquakes on the same terms—as emergencies for Mexico City residents. I go on to describe how technoscientists have worked to make warnings part of Mexican life, with particular attention to the limitations and opportunities that they found in doing so. I end by returning to the story that opens this chapter to consider life with both earthquakes and technoscientific efforts to change the way people experience earthquakes. There will be no disasters in this chapter, no massive earthquakes, and no destruction. Instead, I foreground what happens as we prepare for and attempt to mitigate the worst effects such events might have.

NOT ONLY THE BIG ONES

A substantial portion of my research on this project involved intensive ethnographic participant observation with CIRES, the NGO that developed and maintains Mexico's earthquake early warning system from a pair of busy buttery-yellow buildings on a residential street in Mexico City. Over ten months of visits and interviews with the engineers, technicians, and administrators whose work made earthquake early warning possible, I learned about the complex collaborations undergirding alerts.

Even in my first formal meetings with CIRES director Juan Manuel Espinosa Aranda and his colleagues, in a room given over to maps and blinking system servers or at massive desks in the building's biggest and least-cluttered offices, I learned that very little about earthquakes, or early warning, was as obvious as it may seem. Conducting interviews, sitting in on meetings, and working at a computer terminal while activity swirled around me in the following months only served to support this initial impression. Earthquakes are complicated events, and even CIRES's efforts to focus only on the instrumentation of early warning—siting and maintaining field stations, developing and refining algorithms for seismic analysis, and delivering alerts to a handful of cities across Mexico—was undermined by the complex entailments of system function. Perhaps in ideal cases, where alerts sound and people take cover before massive earthquakes roll in, CIRES's efforts to focus on technology and leave alerting and education to others would be sufficient. The diverse forms of seismicity and experiences with alerts made this impossible, though.

It is common to talk about earthquakes as if they are obvious, focusing on incidents of significant shaking that require people to act to protect themselves. Such events are, after all, what most charismatic earthquake risk mitigation technologies and policies are designed to address. Early warning systems are no different; experts and popular accounts alike often describe how these technologies can detect and then communicate about significant shaking to trigger human or automated action, rather than the kind of alert that I experienced on September 29. Everyday life with and work on early warnings, however, requires considering more than massive quakes and the dangers they pose to human lives. Thinking only about big events means denying empirical complexity, focusing instead on those few cases that fit into a single narrow category. Severe earthquakes are important and dangerous, but understanding life with moving earth and technoscientific intervention requires something more. It is important, instead, to seriously consider the diverse experiences that people may have with the hazard as well as the technology, not just to account for ideal use cases but to understand real ones.

Considering how this technology contributes to contemporary Mexican experiences of seismicity means addressing what it is like to live with an early warning system, which, in turn, requires accounting for the fact that experiences with the warning constitute emergencies regardless of whether any earthquake follows. By emergency, I mean a sudden disruption, something that requires an urgent response and has potentially serious consequences. "'Emergency' is now the primary term for referring to catastrophes, conflicts, and settings for human suffering," sociologist and political theorist Craig Calhoun explains. He understands it to have "rough cognates such as 'disaster' and 'crisis,'" but, as he points out: "Use of the word focuses attention on the immediate event, and not on its causes."[8] The "immediate event," when it comes to early warnings, need not be an earthquake itself. As I felt viscerally in the apartment on Plaza Popocatepetl on September 29, the warning itself is an emergency: Beca, Enrique, and I were surprised. We made decisions—moderately quick ones, although we were still in the building by the time the siren stopped sounding—about what to take with us and what to leave behind.

Although the nature of emergencies varies significantly, scholars and professionals in disaster response and recovery alike have developed certain principles for communicating about hazards and dangerous conditions.[9]

The best modes of dissemination take users into account. The early warning experts I worked with in Mexico often referred to guidelines established by Denis Mileti and John H. Sorensen's extensive review of early warning systems for all manner of phenomena. Good warning messaging should contain "consistent, accurate, and clear information; guidance on what to do; risk locations; and confidence or certainty in tone," and should come from sources that the groups targeted for messaging consider credible.[10] The production of that trustworthy information and guidance should, they stipulate, be calibrated to users and their needs, and offered by multiple sources with as much detail as possible.[11]

In emergencies, it is essential that authorities communicate key information. It is also important for people to be able to find more information about what is happening and why as they need it.[12] People involved in earthquake early warning, including engineers and their client governments, have pursued a variety of strategies for authoritative communication and making further resources available. Early warnings that once went to dedicated receivers and broadcast only on public radio and television now echo from sirens mounted on telephone poles and buzz from smartphone apps. Nonetheless, some issues related to effective rapid broadcast emergency communication remain constant.

Public ability to respond effectively in an emergency is related to the context of the warning—the second crucial element of living with early warning. When we experience warnings, we understand and act on them in the context of how we see the world.[13] Everything, from how we understood the content of warnings themselves to the way we felt about the agencies that circulated them to our preparation for hazards and the way each of us reacts to surprises, had implications for how Beca, Enrique, and I responded to the warning we heard on September 29.[14] After many years of living in Mexico City and a few years of hearing my research stories, Beca and Enrique understood as well as I did that the system could indicate an earthquake but might not be entirely accurate. We took the warning seriously. However, we could have been faster at getting ready and getting out the door. We were not as prepared as we might have been, though for different reasons—Beca was asleep, and she and Enrique were just getting used to the city's new system of loudspeaker-aided warnings. I was out of my element and disoriented by travel.[15]

Even those early warning system administrators who assured me that their responsibilities are confined to instrumentation alone also told me that getting users—even willing ones like me and my friends—to take advantage of alerts was both crucial and challenging in the extreme. Although there is a wealth of investigations on effective strategies for emergency communication available to earthquake early warning system designers, these suggestions are only sometimes useful. Earthquakes happen much more quickly than hazards like floods, hurricanes, or droughts. Insights related to other hazards may simply not be helpful in the case of earthquakes, for which warning time is so tightly constrained. With only seconds to spare even in a best-case scenario, it may be impractical to include the kind of information that Mileti and Sorensen recommend.[16] Communication must happen rapidly enough that people can use the information, and alerts must be brief and clear. Short and easy-to-understand messages are especially important for earthquake early warning, as the window of opportunity to act is so small. Research has also highlighted the advantages of including directions and images, but there is strong evidence that emergency communication should not be considered in isolation. Follow-up messaging and familiarity with protective actions are also crucial.[17]

Communicating earthquake early warnings is one thing, and a difficult one at that. Communicating *about* earthquake early warnings presents its own additional challenges. Mounting a large-scale public education campaign has been beyond the scope of the engineers at CIRES. As engineers would tell me, they were technically trained and their responsibilities were related to their areas of expertise. Nonetheless, it is essential for users to understand how early warning works and how they can use it. CIRES engineers strategically navigated the tension between the needs of their project; the social, financial, and technological resources available to it; and the scope of their expertise.

Communicating about earthquake early warnings has required significant interagency coordination, as I learned in the CIRES offices. There, a great deal of higher-ranking engineers' time was devoted to strategizing about and managing these collaborations. Although this earthquake early warning tool is designed to change the way people experience earthquakes, engineers often discussed how vexing the challenges of public communication could be. As CIRES engineer Antonio Duran reflected candidly to me during one of our long meetings, "The human factor always puts us in our place."[18]

Communication is crucial to the success of risk mitigation projects but was beyond the scope of what the engineers at CIRES could bring to bear on the project. To me, it seemed that this did more than put the engineers "in their place," or keep them humble, but rather caught them in pernicious trap. As SASMEX's only real consistent advocates, they had taken responsibility for the system. However, the system was always failing to meet expectations—in part because its success depended on kinds of social science work that they were neither trained in nor funded to do and required influence in policy and popular culture that they could not access.

DISTRIBUTING ALERTS

While engineers at CIRES were by no means trained in social science, SASMEX has been developed in ways that show some consideration of social dynamics. Take, for example, the problem of spatial warning distribution. In Mexico City, the ground in one place may respond differently to seismic effects than that in another, a phenomenon that has been documented with many seismic microzonification studies.[19] It might seem sensible to some of us to customize early warning systems to each different area, at least before mobile phones began to travel everywhere with people. Sirens could warn those in the sensitive lake regions to earthquakes of lower magnitude than those living on hard rock. But Mexico City is also known for its inequality and rampant clientelism, and many of the wealthy areas of the city are on sensitive soils. This could conceivably lead to a situation where the rich benefited from sirens and the poor did not.[20] For this reason, earthquake warnings have never been disseminated only to certain sites. They have instead been circulated throughout Mexico City and, later, other user communities who contracted with CIRES to use the system primarily by means of broadcast. The CIRES team has asserted that they are primarily responsible for instrumentation and the technical operation of the earthquake early warning system, and that others should take on decisions like where and how to spread alerts. However, while Civil Protection authorities may be involved in making choices like setting alerting thresholds or designating areas to receive alerts, their efforts to educate people about the alert have been limited. Thus, despite their insistence that they focus on technologies rather than the social aspects of alerting, CIRES engineers are nonetheless in a position where they must grapple with how people relate to the technology they develop and maintain.

Before loudspeakers throughout Mexico City first began to howl out *alerta sísmica* in 2015, earthquake early warnings were primarily propagated by television and radio stations. It was a good strategy, and the CIRES team told me that they were happy to have professional communicators passing on the warnings their technology produced, but it fell short in several ways. First, media watchers were simply limited. In 2013, before the sirens, a survey of over thirty-three million Mexicans indicated that most (77.32 percent) watched between one and eight hours of television per week.[21] Second, broadcasters only rarely described how early warning works, leaving people with limited context to understand what an alert might mean for them.

Partnerships with television and radio stations were, in short, not working as well as they might have. CIRES developed other tools to integrate alerting into peoples' lives and do careful outreach in the process. As an alternative to the broadcasting system, wealthy companies, government offices, and schools contracted directly with CIRES to have on-site radio receivers installed in their buildings dedicated to detecting signals from the earthquake early warning system. CIRES complied, developing and maintaining large radio equipment for these clients, often with rooftop antennae carefully arranged to receive signals in dense urban spaces where a line of sight could not be guaranteed. Of the available options, this method offered the most controlled, reliable way for CIRES engineers to get an early warning to people at risk. As of 2014, 403 Personalized Seismic Alert Systems (or SASPERs) receivers had been installed in Mexico City. Schools have these devices, along with banks, government offices, and courtrooms. Though CIRES does not plan to install any new ones,[22] the "cabinets," as they are called around CIRES offices, are still maintained through ongoing cycles of personalized and time-consuming technical attention.[23] Technicians have to visit each one three or four times a year to prevent problems, to ensure that the backup batteries are not slowly draining from electrical interference, and that the technical tools have been cared for.

By interviewing personnel and observing work in CIRES offices, I learned that these visits presented opportunities not only for tune-ups but also for conversation and education.[24] Technicians, primarily young men, would go out daily on maintenance visits or even to set up tests and drills for users. I saw them leave CIRES headquarters with tools in the early morning in cars emblazoned with the NGO's logo, or, if they got in too late to requisition a car for themselves or were headed somewhere without parking, on one of Mexico City's many public transit options. They worked in pairs to fix or relocate parts of the receivers: the loudspeaker, antenna, or the "cabinet" itself. Teams of three or more would go out for each installation. Back at the offices in their workrooms, they explained to me that they moved from one user to another, circulating, testing, connecting, and writing reports. Meanwhile, they would also answer questions and train users, helping them understand the system and how to use early warning technologies. The work that technicians do is, in this way, a key (though often less obvious) part of CIRES's education strategy.

Others working at the NGO use different methods to educate people about how to interpret and respond to warnings. CIRES director Juan Manuel Espinosa Aranda appears regularly in news and print media, explaining warnings and the organization's work. For lower-stakes and less visible work, there is an entire department at CIRES tasked with maintaining outreach efforts. When I visited, this team sat at computers a floor below the technicians.[25] There, they used their skills in computer science, law, and design to make sure that data produced by seismic field stations were accessible to users. They developed blog posts, tweeted, posted on Facebook, and generally monitored SASMEX's media presence. They designed outreach and education materials and sometimes operated programs appropriate for changing messaging strategies. Through these varied mechanisms, the CIRES team advocates for earthquake early warning in public spaces, attempting to make it manageable and comprehensible. They engage with topics of seismic risk mitigation generally and help people make sense of the earthquake early warning system and its uses.

Even though the CIRES team's labor is significant, the technology has not garnered easy acceptance in Mexico City. New support from agencies and collaborations with private alerting firms (discussed further in chapter 4) have brought new users and technologies into relation with SASMEX and its signal. New regulations that require businesses of a certain capacity in Mexico City's high-risk zones to purchase and install some form of radio receiver have caused an increase in the absolute number of users, while the means by which they receive warnings have diversified.[26] Users may encounter an earthquake or warning in different contexts, such as different times of day or in different built environments. Those who hear the warning may be visitors or simply unfamiliar with how to interpret it and what actions they should take. When sirens were first activated in 2015, there were similar misunderstandings. While CIRES technicians try to have face-to-face interactions with the majority of those who own dedicated radio receivers, they can hardly reach all possible users. As different ways of encountering the alerts become more common, CIRES technicians have little say in potentially crucial questions of equipment operation or emergency action. Subscribers to the increasing number of smartphone app notification services may not know how to interpret the messages they receive or respond to them. SASMEX's infrastructure is decentralized, but no organization besides CIRES seems to prioritize supporting new ways to live with earthquakes and earthquake early warning. Despite all the CIRES team's efforts, their work alone has not been sufficient to help people understand and use earthquake early warning.

A study of users of dedicated, CIRES-maintained radios in 2009—before the sirens and smartphone apps—indicated that even for these users, education about the warnings might be lacking, or at least partial.[27] Among the users surveyed, researchers found that people did not accurately understand how earthquakes worked or, more worryingly, how SASMEX worked and the standards against which it should be assessed. Of those who responded, 91 percent regarded the earthquake early warning system as helpful. However, most could not explain the differences between the two forms of warnings they might receive—that preventative sirens might sound for high-risk populations, like school children, at any earth motion above magnitude 5.5 at the source, while the public at large would receive warnings for all earthquakes at magnitudes above 6 that might shake them.[28] Many did not understand that the territorial arrangement of the system's seismic stations would mean that they would get warnings about only some earthquakes, particularly those originating on the west coast of the country.[29] Further, there was the question of what to do with warnings. For all the counsel that CIRES technicians provided, many users claimed they had received no support for developing or implementing evacuation plans. A more recent study has demonstrated that some of these issues are ongoing. In the wake of two 2017 earthquakes, Jaime Santos-Reyes, a systems engineering researcher in Mexico City, showed substantial inconsistency in how people understand the capabilities of SASMEX. More than half of those surveyed indicated that they expected a standard sixty-second warning time for any earth motion.[30] While most respondents reported they were confident that they knew what to do when the warning sounded, many had not found it helpful in a recent earthquake when the system had given them substantial warning before shaking began.[31]

Given that campaigns to help people understand and use the warning have been so limited, it is perhaps not surprising that most people are not quite sure what to do when they hear the simple earthquake warning sound. The general tasks of sharing information related to early warning and advocating for its use require help from people with different connections and training than most of CIRES's staff have.

INSTRUMENTATION AND ITS ADVOCATES

In ordinary conversations around their offices, CIRES engineers and technicians told me that their focus was on instrumentation, not the complex relationships between social practices and environmental conditions that SASMEX was designed to fundamentally change. For all they understood how important it was to their project to consider how people might really live with both warnings and earthquakes, engaging with these issues fully was outside of their capabilities. They were trained in electrical engineering and computer science, and specialized in seismological technology. They could develop equipment to register, analyze, and broadcast warnings about quakes, and advocate for it. They could stray from their commitment to instrumentation to train those few users who had their own dedicated in-building SASMEX receiver and alarm systems, or advocate for the system in popular media. But the system the CIRES team operated would never work on their efforts alone.

Instead, CIRES relied on partnerships, which could be challenging and inconsistent. Their nature could change with the political winds and the agendas of leaders who rotated through agencies, but they could also benefit from shared interests. For example, the CIRES team's work to reduce risk among vulnerable populations could bring their project in alignment with other organizations. The way that CIRES focused on school safety provides an example, allowing them not only to take on public education and outreach by other means but also to enroll motivated allies with crucial expertise to consider what life with warnings could be like. Schools provided an opportunity to protect high-priority and vulnerable populations and make educational interventions with potentially broad effects. For this reason, schoolchildren became an important target user population for the warning as the system developed in the 1990s.[32]

Protecting children was a shared value for collaborators in CIRES and Mexican government agencies. The stakes of protecting children are high, and keeping them safe was an important goal for risk management. More than that, though, children represented a particularly excellent focal population for efforts to change how people living with earthquakes used new alerting technology. Children could learn, and classrooms could provide an environment where they could be drilled and that knowledge reinforced. They could then serve as vectors, bringing what they learned home from school and teaching it to their families, and could grow up to teach their own children.

The decision to focus on protecting children made it possible to mobilize collaborators with different kinds of expertise. Elia Arjonilla Cuenca, a sociologist who began to study and design earthquake safety in Mexican schools after the 1985 earthquake, was one such collaborator. Her presence in the earthquake risk-mitigation community and her work studying and advocating for earthquake early warning from an early stage shows just how critical external support has been for SASMEX.

It was CIRES engineers who first introduced me to Arjonilla. Now retired, she is still active and thoughtful about her work on earthquake safety. Over coffee in the beautiful Museo Dolores Olmedo, the small, elegant woman spoke about her work frankly. Arjonilla explained how her social background as well as her education facilitated her work on earthquake risk and risk mitigation. With family members and friends who studied earthquake monitoring, she was part of the seismic community by association. Arjonilla herself had received training in sociology and public health, a combination that made her ready to organize studies and advocate for public safety. This background positioned her well to participate in technical and political talk, from conversations over meals to community events in the wake of the 1985 quake and the reevaluation of safety procedures that followed.

Arjonilla described to me how she began to enter scientific gatherings. After the earthquake, there were visitors from around the world. Concepts and people circulating through Mexico City from Japan were particularly exciting. She went to all the seminars. "I wrote and wrote and wrote," she told me, "and I asked questions."[33] It was important to her to be involved, but she knew she would seem out of place in a highly technical and male-dominated space.

She described how deliberately she went about introducing herself in such gatherings. "I'm overjoyed to be here with you today. I work in seismic education," she recited for me. That was her line. She knew that she was not like the other people present. She could not hide it, so she did not try. Instead, she made sure to tell people about the parts of her work she knew they would care about.

> It was like opening doors! Once you identify yourself as a nontechnical person with a very important job . . . well, they were very interested in building seismic education at that time. I would say "I work in seismic education, and I'd like to know . . ." well, whatever I wanted to know. Their responses would be better, their attitudes better, so much better. And if your question wasn't answered by the speaker, maybe someone from the audience would catch up with you at the exit and say, "That's fascinating. Where are you based? What are you doing? I'll give you my card."[34]

In 1985, when the disastrous quake struck, Arjonilla had children enrolled at Colegio Madrid, a private school that children attend from kindergarten until they are old enough to attend university. The Colegio is in Mexico City near Xochimilco and on seismically sensitive soil. Her extended family was soon deeply involved in the seismic community working busily at the National Autonomous University of Mexico and within Mexico City's government to understand the seismic territory on which they lived in new ways. The quakes had damaged her children's school quite badly, and some parents of the Colegio's three thousand students decided to take part in a project called the Parents' Volunteer Safety Commission. She used the contacts she made to advocate for new safety procedures.

Arjonilla left her position in medical sociology to take that post-earthquake volunteer job along with other parents. Over time, she took on more responsibility. Part of her work involved looking at the earthquake procedures in places elsewhere and developing plans to implement the best of them as a safety program for the school. In 1986, Arjonilla and another parent worked with US-based disaster researchers on an English-language report on the Colegio's recovery.[35] She remembers picking up the technical language of seismicity quickly from conversations with family and friends, and the meetings she attended. She paired it with the professional knowledge of risk prevention that she already had and brought both to bear on her interest in school safety.

When other parents scaled down their involvement, Arjonilla kept working. What she had been doing as a volunteer became paying work. "You know what they called me?" she laughed. "The Earthquake Girl." Her gender and her youth made her stand out, but she slowly gained influence. She continued her work at Colegio Madrid, contributing to safety assessment and training teachers, students, and staff. Then she became a consultant with other schools, the education ministry, Mexico City's government, and then eventually with CENAPRED, the National Center for Disaster Prevention.

Another sizeable earthquake struck the city in 1995, and Arjonilla was able to compare responses in a school that used SASMEX with one that did not. When she interviewed students who had experienced the same degree of shaking in similar structures, she found that, overall, SASMEX had helped. The school that had an early warning followed safety procedures without difficulties; the earthquake emergency produced less tension and disruption than at the school that lacked the system.

While Arjonilla broadly supported expanding earthquake early warning systems, she also argued that, from her observations, integrating earthquake early warning into practice in a sensible way would require "solid planning and preparation on the part of the community."[36] She was, in other words, an engaged collaborator and critical of the lack of thoughtful public communication involved in earthquake early warning rollout. She wrote that CIRES was "expert in questions of monitoring, detection, and warning, in merely technical terms," but it did not have "the perspective of other disciplines necessary to promote an optimal collective response."[37] Her vision of an "optimal collective response" would include large-scale education and outreach, and better emergency training for all.

Despite Arjonilla's advocacy and that of many others, not all schools were able to use earthquake early warnings. It is difficult to tell how many did. Out of 5,500 schools in Mexico City in 2009, only seventy-six had their own SASPER cabinets and associated contracts with CIRES.[38] It is likely that many more had less expensive emergency radios tuned to the service. As today, with increasing options for early warning transmission, it is challenging to document and evaluate warning use, even with the reporting required by Civil Protection. In an effort to overcome this limited use, in 2010 and 2011, Mexico City and federal governments purchased 88,000 small emergency radios for distribution to users around the nation.[39] Nearly forty thousand were destined for primary and secondary schools in Mexico City alone. Although there are records of the city and national governments purchasing this equipment, journalists have suggested that many of these devices did not make it to their designated recipients.[40] Those schools that were unable to buy SASPERs, to purchase their own small emergency radios or to receive government-sponsored dedicated radios may have found other ways to access the warnings. While this policy was outlined in a memo for public schools in 1995, its rate of implementation is unclear.[41]

Arjonilla's experience was unique, animated by her passion for safety reform. Gracefully yet insistently, she made a space for herself and her insights in a field dominated by physical scientists and engineers. She became a volunteer, an educator, and an "Earthquake Girl" to bring socio-logical insights to risk mitigation work. She was able to help identify the challenges that would plague SASMEX, but this kind of attention and sup-port combined with the CIRES team's creative commitment alone could not mobilize the kinds of large-scale resources that integrating the earth-quake early warning system into Mexican awareness required.

AN EARTHQUAKE EMERGENCY IN 2015

In 2015, it took only seconds after the siren stopped for my friend Beca to grow tired of holding her squirming dogs outside her door. She knew there would be people outside, talking through their shared experience. "Take the keys and tell us what you find out," she told me, and, sensibly, went back into her apartment with Enrique and the dogs. They shut the door behind them.

Down on the plaza in front of their house, I found a handful of people talking. They were gathered near the central fountain in little groups, away from buildings and bathed in the green glow of streetlights filtering through the canopy of trees (see figure 2.2). Against the white of the fountain's arches, their outlines were visible, even blind as I was without my contact lenses.

"I suppose there has been an earthquake," a young woman in pajamas told me. "When we came down, the water in the fountain was rippling." Another was more skeptical. "It's rippling," she said, adjusting her short blue dress, "from the wind."[42] A magnitude 4.8 earthquake originating in

FIGURE 2.2
The author and Beca Gonzalez walk past the fountain in the middle of Plaza Popocatepetl at a more leisurely moment. *Source*: Stephen C. Rea (2018).

Guerrero might just have made the water in the fountain on top of one of
the most sensitive soils in the city ripple. But there was also a light breeze
that night.

When I returned to the apartment, I scanned Twitter and Facebook messages for more responses, reading about how people had become caught
up in SASMEX. No single hashtag surfaced that evening as a locus for commentary, but I found many messages in response to tweets from official
business and NGO accounts affiliated with earthquake early warning, visible to anyone watching the lively feeds. I saw frustration, commiseration,

and confusion. Many referenced the scare of the warning and the bodily responses it had engendered in them. "Fuckin' fright!!!" one wrote, appending a photo of an adorably horrified-looking child. "I just about died of a heart attack in my baby's arms," another posted.[43] The tweets may have been hyperbolic, but they reflected a community dealing with stress. Others posted pictures of bloodshot eyes, frightened cartoon characters in bed, or of bread rolls they would eat (*"pa'l susto"*).[44]

People were frightened of seismicity and understood encounters with this fear to have consequences, even without any physical shaking. According to the Ministry of Civil Protection in Mexico City, there were no injuries or damages that could be attributed to the quake, but some residents took the experience itself seriously.[45] Commenters understood that a warning might remind some of old earthquake trauma, or could itself cause new emotional and physical harm.[46] Although social media commentary is not necessarily an indication of earnest response to an event, these posts demonstrate that the warning was broadly an unpleasant experience, and a matter of serious concern for Mexicans.

Even on the night of the alert, those who had received the warning debated the long-term effects of "false alarms." There was some discussion of preparation and training on Twitter. "It can work like a drill," wrote one person. "Every false warning should be an opportunity to practice a successful evacuation in case an earthquake happens. I don't know why they just complain!"[47] The sentiment was not uncommon. The warning, here, could be considered part of a training regimen for speed and organization in evacuation, an opportunity to rehearse for a more dangerous earthquake. Recipients of a warning run through the physical steps and experience the tension of an unexpected experience so that, in the future, they are more likely to know what to do and, moreover, confident in their actions. Drills, rehearsals, or practices of this sort have become ways that people make themselves ready for an otherwise unpredictable emergency, and ready to avoid the worst of its potential effects. Risk mitigation experts like Arjonilla have, for years, urged the public to use false alarms as opportunities to train for the real thing. In practice, however, the issue of the earthquake-yet-to-come surfaced in other ways.

"They need to recalibrate the system," Enrique told me over coffee, the morning after the loudspeakers had sent us rushing out of the apartment, "or no one will trust the warning." A post on the Facebook wall of the earthquake early warning mobile phone application SkyAlert offered a similar sentiment. It read: "Me, I left the house when I heard the alarm. I think it's unwise for the government and that company to make panic with a tremor of 3 degrees. It reminds me of the joke about 'crying wolf.' After all this, they'll lose credibility." The "joke" referenced a fable attributed to Aesop, sometimes titled "The Boy Who Cried Wolf." The fable involves a boy tasked with guarding sheep. The boy issues false alarms about a threatening wolf and loses credibility as a result. In some versions, the false alarms and the subsequent inability of the shepherd boy to find help when he needs it costs him his flock of sheep. In others, he pays with his own life.[48]

An article in the Spanish newspaper *El Pais* recounted similar concerns. A warning without a perceptible earthquake following—like the earthquake emergency I experienced with Beca and Enrique on September 29—might have implications for a large quake that is likely to occur sometime soon. The emergency could blossom into disaster if Mexico City experiences a large earthquake and people neglect early warnings; they could be injured or killed because they expect another misfire.[49] Previous experiences could lead to better awareness and preparation or, conversely, might result in avoidance or a naively optimistic assessment of safety.[50]

Emergency responses, especially responses to quick-moving hazards like earthquakes, generally work best when they are trained into the body and come to entail simple and unconfused physical reactions to certain stimuli. This is one of the reasons that drilling and similar exercises have become so important in emergency preparation. They provide opportunities to rehearse crisis, to condition embodied and mental responses, and encounter safety tools and techniques intimately. In this context, a warning like the one on September 29 could interfere with effective responses to future earthquake early warnings. Or, just as easily, it could facilitate them.

Public earthquake early warnings use technoscientific means to give users a chance for speedy response in case of a quake. If people are not interested in taking that action, then the advantage that early warnings provide decreases tremendously. With the loudspeaker broadcast, Mexico City provided a new way for people to use SASMEX. Juan Manuel Espinosa Aranda was interviewed about it by a journalist. He commented, "[The warning] was warranted because we do not control the phenomenon, because we have no certainty that all the structures of the city are safe."[51] Broadcasting the warning could make a tremendous difference for Mexico City residents in dangerous places.

CIRES engineers continue to tweak the system. In an earthquake on March 23, 2016, of a similar magnitude to the one that triggered the warnings discussed here, no warning went out over the loudspeakers.[52] Refinements can only do so much, however, to control for the many ways that users experience a technoscientific intervention in their lives. Even the moderately sized earthquakes for which Mexico City has chosen to alert its population might be felt strongly in the areas of the city with the most sensitive soil, and less in those built on hard rock. Any warning will intervene in the lives of Mexico City residents; this general broadcast is particularly promising because it is available to so many. This remains true even as smartphone use rises; apps may be popular, but they face certain technical limitations. The ways the warnings relate meaningfully to the material world, however, is a matter of diverse practice rather than an automatic effect of the technology.

THE PRINCIPAL SUSPECT

Enrique messaged me soon after I left Mexico City to joke about my obvious fascination with the events that continued to unfold during and after my visit. "You are now the principal suspect of producing earthquakes! I am sure you are playing with the thingie to see how we, your lab rats, react!"[53] Enrique's comment was self-consciously absurd, but I could not deny how useful for my research it was that an alert should happen when I was around to experience and analyze it. The importance of empirical social scientific research on how these events unfold in context cannot be overstated, especially when so many accounts of early warning systems focus on isolated warnings and evaluate technical successes and failures.

In this chapter, I have suggested that we reorder and expand how we conceptualize earthquake early warning to include more diverse seismic and warning events—events that may happen more often than those who are not playing host to inquisitive anthropologists realize. People do not simply engage with earthquake early warnings as precursors to earthquakes; rather, they live with warnings as well as earthquakes. Understanding that allows us to think in greater scope than the simple techno-optimistic approaches to seismic disaster prevention, which suggest that earthquake early warning technologies alone will save lives and transform users' experiences of seismicity simply, without causing other challenges.

Considering the different kinds of experiences that a warning might entail means addressing challenges that include: the ongoing threat of a violent earthquake; the promise of new opportunities to evacuate or take cover before the danger; the experience of a late-night shock; and the real capabilities and limitations of technical interventions into life with complex, unpredictable, ongoing seismicity. Earthquake early warning is not simply a communicative act but one part of a complex of meaningful events around seismicity in Mexico City.

II Risk Mitigation Technology

3   A TECHNOLOGY TO
MITIGATE RISK

The Third International Conference on Earthquake Early Warning in 2014 was a packed event, with attendees from around the world. It opened with a talk about the urgent work that drew conference participants together (see figure 3.1). The first speaker, UC Berkeley seismologist Richard Allen, had long been a strong advocate of earthquake early warning systems. He saw great potential for these technologies, even in places that sat closer to faults than Mexico City. No longer would earthquakes take people by surprise. Seconds of warning before earthquakes struck would open up new opportunities to protect vulnerable people. Whatever promises earthquake early warning offered, Allen reminded all of us that these systems would only work if they were implemented. Allen had, at that point, yet to secure the support necessary to launch a truly public earthquake early warning system in California, despite the state's well-known seismicity and thriving tech industry. He was working hard to do so, though, and his work to rally interested parties and energize collaborators used Mexico City as a touchstone. He explained:

> Mexico City has a warning system, built after ten thousand people were killed in 1985. The question is, therefore, what would it take to build a public earthquake early warning system in the United States?[1]

Over three days in Berkeley, we heard about a variety of sensor instruments, data processing algorithms, and uses associated with earthquake early warning systems around the world. All earthquake early warning system design is about the relationship between earthquakes and human well-being. Roughly one hundred of us had gathered in a wood-paneled auditorium in Berkeley to

FIGURE 3.1
Richard Allen introduces a distinguished panel of Californian politicians, policy makers, and seismic policy advocates (from left to right, those seated are Jerry Hill, Gavin Newsom, Alex Padilla, Mark Ghilarducci, Ed Lee, and Lucy Jones). *Source*: *Berkeley News*, photographer Peg Skorpinski (2014).

learn about cutting-edge earthquake early warning technology. The 2014 conference brought primarily scientists and engineers but also emergency managers, policy makers, and inquisitive anthropologists together from universities, businesses, NGOs, and government offices across East Asia, western Europe, and North America.[2] Teams of experts gave talks, one after another. At regular intervals the talks broke for coffee and snacks, and to give us time to peruse a small set of posters and draw each other aside for focused conversations in the hallway or in the California sunshine. There was professional diversity here, but a few trends too: discussion tended to be grounded in the geophysics and engineering of these systems; it seemed built on commitment to addressing seismicity as a danger to human well-being; and it showcased a shared confidence that earthquake early warning made sense as part of such efforts.

Most of the people attending the conference had come to talk about what they saw as technical aspects of these systems: high-tech sensors, data analysis, and communication techniques to change the way people experienced earthquakes. Allen's introduction reflected the way that, for all that these technoscientists outlined different subject matter, many thought that earthquake early warning would save lives, property, or money. The people at the conference, drawn together by this common interest, generally seemed optimistic about early warning's utility. Most, however, did not discuss this utility in any substantial detail. Their presentations often mirrored Allen's by describing various early warning technologies as the benefits were obvious and positive impacts inevitable, just a matter of some issues that still needed to be worked out. These potential benefits may not have been fully articulated, but they seemed to be persuasive to many—I only had to look at the who's who of earthquake risk mitigation and California politics on the stage[3] or listen to the ways people spoke about their projects to find evidence of how much promise the people in this community thought that earthquake early warning had.

For all that ShakeAlert was of primary concern to conference hosts and local attendees, there was a good reason that Allen referred to Mexico and its SASMEX in his opening remarks. Mexico's public system had been the first of its kind; in comparison, most of the technologies being described at the conference were in their infancy.[4] SASMEX belongs in any introduction to early warning and its opportunities.

I was attending the conference to learn about approaches to early warning outside of Mexico so as to better understand the Mexican earthquake risk mitigation project within this international context. Two men from CIRES, the NGO responsible for developing and maintaining SASMEX, were there too, though their priorities were less about understanding their own projects and more about showcasing them. Juan Manuel Espinosa Aranda had directed CIRES since it was formed in 1989 and seemed to know nearly everyone at the event. Armando Cuéllar was meeting many for the first time. He was a thoughtful man in his thirties with a PhD in geophysics nearly completed and was responsible for representing CIRES as the head of the organization's Public Outreach Department. The Mexican earthquake early warning system that they came to speak about seemed well known within this community.

In his comments at the 2014 meeting, Allen moved on quickly from the solemn reminder of the disastrous Mexico City earthquake and the development of SASMEX. For Allen, as for many of the others listening, this was a way of framing technical innovations and drive toward implementation.[5] Sitting beside Cuéllar and Espinosa Aranda, who I knew to be working around the clock to refine SASMEX and advocate for it, this way of characterizing earthquake early warning's development seemed detached from reality. It sounded easy. By 2014, with several years of research under my belt, "easy" was not a word I would ever use to describe SASMEX's development and implementation. In this chapter, I address how technoscientists in this community describe earthquake early warning systems' benefits— and foreground its promises while neglecting the challenging aspects of real-world implementation.

To do so, I draw on my experiences as a participant observer at conferences like this one as well as reflections on popular media, scholarship, and archival documents. I consider prevalent ideas about risk mitigation that circulate in international communities and Mexico, and I argue that the way that this international community understands the promises of earthquake early warning needs to be understood in the context of contemporary ideas about how hazardous environments threaten human life and well-being. Further, these ideas easily support techno-optimistic trust in earthquake early warning to change relationships between environments and society, without careful attention to issues of specific use. In this chapter, I introduce the ideas about seismic hazards and people at risk that inform earthquake early warning design. I historicize the development of these technologies. Finally, I describe these global trends in risk mitigation in the particular forms that they take in Mexico. In this way, I show how earthquake early warning technologies make sense to so many as a tool to mitigate risk in hazardous environments, and I situate SASMEX in the context of these logics.

HAZARDOUS EARTHQUAKES AND VULNERABLE PEOPLE

For most, if not all, of the technoscientists I sat with in Berkeley (and many at other conferences before and since), the term "earthquake" is a way to describe how the ground shakes. Earthquakes are generally caused[6] by constant and subtle shifts in our planet's crust.[7] This may happen for a number of reasons—the earth is, after all, a thermodynamic system of pressure and motion. Tectonic plates are always moving in different directions, over, under, and against each other, but immense friction also keeps them from slipping while stress builds up at faults within plates and at their interfaces. Eventually, the stress is great enough that rock slips against rock, releasing seismic energy, which travels through the earth at roughly the speed of sound according to physical effects of density, rigidity, resonance, and refraction. Earthquakes move through various materials differently, with a variety of measurable effects evident in the waters, soils, and built environments that respond to and propagate quakes.[8]

By the late nineteenth century, a number of mechanical tools had been developed for understanding earthquakes, most of which were associated with research programs in Europe, East Asia, and, to a lesser extent, the United States.[9] When contemporary earthquake early warning systems register quake motion and describe it in terms of the magnitude of energy that a quake releases or in relation to wave frequency or ground acceleration, they are relying on tools and using language that technoscientists around the world share.[10] However, the way we describe earthquakes remains contingent on other issues—including why we understand earthquakes to be worth describing in the first place.[11]

For those gathered at the Berkeley conference in 2014, earthquakes were not just geophysically fascinating subjects for investigation. They were also hazards. This distinction is both crucial and deceptively simple. For those involved with emergency management and disaster prevention, it is not only necessary to grapple with the physical characteristics of the unpredictable and ongoing movement of the earth's crust but also to consider their implications, that is, the destruction that such motions cause. When earthquakes are not just seismic phenomena but also hazards, describing them means more than talking about faults, waves, or soils; the ways that they threaten people also become key aspects of their definition. How exactly the risk that earthquakes pose should be described is a crucial topic of attention—and debate—for those who build and operate earthquake early warning systems.

In physical terms, an earthquake with a larger magnitude tends to be more dangerous than a smaller one. It shakes things harder. If it originates near a population center, observers are likely to experience more intense shaking, in the language of seismology, assessed by the effects it has on the built environment. But when a quake starts deep underground or far from human habitation, that energy may dissipate and prevent it from becoming a significant threat to people. Certain soils may be more or less sensitive to seismic motion, too. Emergency managers and disaster prevention specialists regard magnitude, distance, and soil qualities as insufficient for describing how earthquakes put people at risk. Other issues are far more important than those.[12]

The United Nations Office of Disaster Risk Reduction and the Centre for Research on the Epidemiology of Disasters address the issue on a national scale, in relation to both physical and social conditions. These agencies calculate that almost 500,000 people were killed by earthquakes between 1996 and 2015,[13] and nearly 126 million affected in other ways—injured, left homeless, or in need of emergency assistance of some kind. People in low- and high-income nations alike experienced similar levels of shaking, but the consequences of the magnitude 7 earthquake that struck Haiti in 2010 were devastating compared to the 7.1 quake that shook Aotearoa, New Zealand that same year. The Haitian earthquake had a hypocenter thirteen kilometers underground, just twenty-five kilometers west of the capital city of Port-au-Prince. New Zealand's was similarly close to a city— just ten kilometers underground and forty kilometers from Christchurch. But while the Haitian government's official tally shows that 316,000 people lost their lives in Haiti—over half of those killed by earthquakes between 1996 and 2015 died then and there[14]—no one died in New Zealand at all.[15] According to the reporting agencies, this terrible difference can be attributed to the comparative wealth of people living in these places, the condition of the built environment, and the involvement and effectiveness of government.[16] These inequities extend far beyond the immediate impact of an earthquake event and into recovery and redevelopment.[17]

When people who study earthquakes call them "hazards," they are alluding to this relationship between environment and society. They mean that an earthquake might cause harm to people but that this harm is not inevitable from a purely geophysical perspective. An earthquake may cause harm, or not, depending on factors that have nothing to do with its physical characteristics. Making sharp distinctions between a hazard and a disaster has been conceptually important for many involved in earthquake early warning, and for people who work in emergency management and disaster prevention more generally. Popular culture and colloquial conversation in Anglophone and Hispanohablante worlds alike are full of casual references to earthquakes as "disasters" by definition—natural disasters, to be precise. To name earthquakes "disasters," however, is to confuse one component of a complex set of events and conditions with the cumulative effects of all of them put together. Not all earth motions have to be disastrous. Social scientists have been documenting patterns in disaster effects for many years, and the idea that it takes more than an earthquake to make a disaster is well established in policy.[18] As historian Scott Knowles has written, we live in "an era of disaster 'adaptation,'" and this has consequences for how we

make plans to minimize disasters' effects. Institutions like university programs and emergency management centers have encouraged these trends.[19]
It has become commonplace to assert that disasters are "not natural"[20] and
can, consequentially, be prevented.[21] The notion that the dangers of seismic disasters might be avoided if only the designers of built environments
made different choices is powerful today.[22]

This mode of thinking is sometimes quantified to integrate it into technical
systems and mathematical reasoning. "Risk is a product of hazard and vulnerability" becomes "risk is hazard multiplied by vulnerability" or "risk=hazard
x vulnerability."[23] This relationship among risk, hazard, and vulnerability is
one way to describe and compare the dangers that one earthquake might
pose to different communities. The formula also serves as a purely rhetorical
device to discuss how social conditions and human action might influence
the likelihood of a nasty outcome.[24] When "risk=hazard x vulnerability,"
attention to social conditions becomes key to diminishing the effect of hazards. This is not the same as a call for greater justice to equalize distribution
of suffering, but it may direct our attention in ways that are foundational
for such claims.

Many people referenced this relationship between risk, hazard, and vulnerability throughout my ethnographic fieldwork. It came up in interviews,
lectures, and archival documents. Sometimes these experts used it to conduct
a practical evaluation of quantified hazards, risks, and vulnerabilities—and
in those cases, it did tend to bear the mathematical signs of "x" and "=" as
marks of their straightforward, mutual dependence. At other times, the for-
mulation simply referred to a collection of issues, each separately defined,
that have effects on each other. Often, it served as an argument for considering ongoing social conditions along with other factors when discussing the propensity for disaster.[25] Operationalized or not, the equation has
marked real disaster prevention and risk management efforts. Since earthquakes cannot be stopped, something about social life would have to be
changed to reduce risk.

The equation circulated in UN documents and disaster policy discourse,
notably in a UN Disaster Relief Coordinator meeting of multidisciplinary
experts in 1979, describing an effort to standardize language around disaster.[26] "Risk" was defined in this report in terms of expected degree of loss,
and "hazards" as probability of an event within a given time.[27] The report
authors proposed that, in the context of the equation, vulnerability could

be assessed in scale from 0, no effect, to 1, "total loss."[28] At that point, minimizing human vulnerability could be conceptualized as a simple matter making changes in the built environment.[29]

Addressing vulnerability to hazards, however, has come to mean more than changing the built environment. Today it might incorporate addressing complexes of ongoing social practices, capacities, relations, and resources that facilitate exposure or interfere with resilience. In the latter half of the twentieth century, experts began to consider this "expected loss" a matter of how people lived, not just where they lived.[30] In recent years, the United Nations program has increasingly asserted the importance of issues related to poverty and institutional capacity.[31] While some challenge the idea that "disasters" begin with hazardous events when radical inequality and poverty already characterize so many lives,[32] others find it easier to acknowledge that social issues have important effects but focus their attention on physical ones.[33] These insights are borne out in diverse research, and guide efforts to describe and model the physical and social factors of disaster impact and long term recovery.[34] All these currents exist in the context of broader debates about environmental justice and whether to emphasize a community's existing capacities, including opportunities to build resilience, or to instead highlight a community's deficits, limitations, and needs.[35]

These ways of understanding hazardous environments and people at risk have had significant consequences for risk mitigation system design. Earthquake early warning systems are primarily designed in light of physical vulnerability, but the systems concern social vulnerabilities, too. Systemic poverty, after all, has consequences for how buildings are constructed and how people rebuild their lives after experiencing a disaster. Even if the Mexico City built environment never became safer, however, and its significant problems with poverty remained unalleviated, the earthquake early warning system could help in the moment of an earthquake.[36]

DEVELOPING ESSENTIAL INVESTMENTS

The UN Office for Disaster Risk Reduction has called early warning systems "essential investments that protect and save lives, property and livelihoods, [and] contribute to the sustainability of development." Warnings are derived from technologically mediated ways of understanding, analyzing, and communicating information about issues like air quality, bombings, tsunamis, extreme weather, floods, fires, and mudslides.[37] In general, these systems work best in places where dangerous events are likely but are not precisely predictable.[38]

In the most basic sense, warnings are meant to make it possible for people to learn about dangerous events as they begin and act based on this knowledge. While any warning is challenging to generate, earthquake early warnings are particularly troublesome.[39] Earthquakes are quick, and unlike floods, they have no reliable precursors.[40] An earthquake can only be detected after it starts. The development of a useful warning under these conditions is no easy task. Earthquake early warning nonetheless captures the imagination. The idea of making new kinds of knowledge about earthquakes available fascinated people long before Mexican engineers started building a system that could do it.

Contemporary proponents of earthquake early warning systems cite a shared ancestor in the nineteenth century, long before the approaches to risk, hazard, and vulnerability outlined above became explicit rationalities for disaster risk reduction.[41] This nineteenth-century plan is the earliest on record for automatically registering, analyzing, and communicating information about the earth's motion.[42] It appeared in the *San Francisco Chronicle*'s Daily Evening Bulletin on November 3, 1868. There, in the middle of a dense page of news, commentaries, and advertisements (just above an announcement about the arrival of a steamer from San Luis Obispo and just below a long and involved account of a Republican political event), a medical doctor named John Cooper described what he called an "Earthquake Indicator" which offered "the means, perhaps, of saving thousands of lives."

This earthquake early warning plan, like those in use today, makes sense in the context of certain conditions and ideas about both the unsteady earth and the people that it might put at risk. As a Californian, Cooper had probably experienced his share of ground motion. He had also been reading up on earthquakes, a matter of popular fascination at the time.[43] For him, an earthquake was "a fearful convulsion" and "a wave motion of the surface of the earth." He based his understandings on the writings of J. B. Trask, a physician who also served as the first unofficial state geologist of California.[44] If, Cooper wrote, an earthquake started far enough away from San Francisco, his system could notify residents before it reached the city. Cooper was proposing a new configuration of what historian Megan Finn has called "disaster informatics"—an information infrastructure that might have ongoing consequences for the experience of earthquake emergencies.[45]

It was clear to Cooper that such a system would need to be automatic and responsive to earthquakes that might happen at any hour of the day or night. It should not rely entirely on human discretion, either. Telegraph operators might be unreliable; they might not be ready to ring the bell when needed, or they might ring it too often. Instead, he proposed a system based on electricity that included a networked "mechanical contrivance" arranged between ten and one hundred miles outside of San Francisco. If a wave crossing the surface of the earth "rose high enough" to set off the mechanism, it could start an alarm bell ringing near the center of the city.

That was Cooper's system as proposed. A successful execution of his plan would require more than just a "mechanical contrivance," however. Crucial to Cooper's scheme was the idea that people—at least, any hearing person in range of the bell, who understood what the sound meant—would know when an earthquake was imminent. They could produce and share knowledge about not just what was happening across substantial distance but what was occurring in the earth itself.[46]

Cooper's ideas about an automated system were part of a broader trend in earth science: a turn toward mechanical sensing. While human senses were, and are, necessary for detecting and parsing the effects of earth motion, mechanical tools were becoming increasingly popular.[47] Such sensory technologies have formatted new ways of knowing the complex geophysical systems that comprise the earth we experience. STS scholar Paul Edwards has described this knowledge as a matter of integrating diverse data collection strategies, analytic techniques, and models to make data global. But this globalism is, itself, infrastructural. Efforts to sense and model earth systems are not just a matter of knowledge production, but they also frame new ideas about what might be possible. The ways we want to understand the earth—indeed, how we expect to know planetary systems and motions and use this knowledge—inform the increasing popularity of environmental monitoring projects.[48]

Early warning systems should be considered just one kind of environmental monitoring project. These projects take many forms that consider and seek to transform the way people live with earth systems today.[49] They do not quite provide the kind of integrated models that Edwards wrote about. Each earthquake early warning system is discrete, incorporating its own sensors and site-specific analytic processes and mathematical models. All share a few basic features. They incorporate automated processes. Some make an earthquake distinguishable from background ground motion, such as that created by a passing vehicle.[50] Others make it possible to estimate an earthquake's magnitude just as it begins to shake.[51] They are networked to communications systems and trigger a variety of different actions depending on both social and technical possibilities. Earthquake early warnings slow trains, send elevators to the ground floor of tall buildings, and raise bridges. They trigger sirens and secondary alerting systems, like smartphone apps.

The first earthquake early warning systems were implemented nearly one hundred years after Cooper suggested the idea. Mexico and Japan were the first nations to develop earthquake early warning technologies, but these tools have also been implemented in many places in western Europe, as well as Romania and China, though not always for public use so much as industrial automatic functions.[52] Systems are currently under development in Chile and Nepal, and, of course, the west coast of the United States.

For all that this tool has become more practical, few sites for earthquake early warning have the affordances that Mexico does; with its rich, populous, and powerful city located at a distance from quakes that would allow warnings to be particularly useful, it has provided an ideal context for earthquake early warning adoption. Critics note that warning systems have very little time to alert for shallow earthquakes that originate close to population centers. If quakes are strong enough to be concerning, their P-waves may provide the most effective early warnings.[53]

These debates make it evident that there are significant differences of opinion regarding the nature of a productive warning within the community of technoscientists who design and debate earthquake early warning systems. Far from a singular kind of technoscientific undertaking, then, early warning systems represent an opportunity to change how people experience earthquakes, potentially in a variety of ways. In Mexico, this technology provided a chance to improve physical safety in a new way that could work in concert with other ideas, programs, and priorities.

PART OF A SEISMIC CULTURE

Talks, posters, and even conversations between sessions of the 2014 conference were often related to sensor technologies or algorithms. However, the presence of fire and emergency service workers in the audience meant that I could never quite forget that earthquake early warning was not just a matter for technoscientists or that speedy data processing would not necessarily mean a successful warning. Pulling off effective warnings required technoscientists to collaborate with others and integrate their systems with institutions and relationships in context. Local coordination is crucial to how and whether earthquake early warning functions. It is never only a matter of integrating computer programs, connecting wires, or even making processes coherent across organizations. It is also a matter of the conceptual integration of priorities and projects in their contexts.

CIRES is officially tasked with developing and maintaining SASMEX's instruments. They look to emergency managers in Mexico's National Civil Protection System to integrate their systems with other technical and social infrastructures. Civil Protection agencies are charged to "protect as well as preserve the individual and society."[54] This work makes Civil Protection part of what researchers call a security apparatus, that is, the collection of semi-coordinated systems and practices that inform how people and institutions orient toward danger and safety.[55]

Civil Protection coordinates emergency action, post-disaster recovery, and attempts to prevent disasters in the first place. Although all thirty-one Mexican states and Mexico City have Civil Protection offices integrated into their governance structures,[56] the agency has been institutionalized differently everywhere. It occupies a variety of positions in governmental hierarchies throughout the nation.[57] While Civil Protection state offices have legal responsibility for resource coordination, these municipal governments have primary responsibility for emergency response.

In Oaxaca and Guerrero, Civil Protection officials gave me the same glossy, stapled booklets I often received in interactions with Civil Protection officials—documents developed centrally to explain one hazard or another, and how it might be prevented from becoming a real risk, selected particularly for those issues most pertinent in each state. Oaxaca officials also shared an internally produced document about Oaxacan earthquakes. I visited the SASMEX servers in the back of the Civil Protection offices, enclosed by dark glass panels beside a screen displaying the status of the loudspeakers that broadcast earthquake sirens through the city, and I saw earthquake early warning as just one effort among many. I also witnessed the real stuff of emergency rescue on display: motorcycles receiving maintenance and an organized mix of departmental and various staffers' personal tools on shelves and along walls, ready for emergency evacuation or first-aid in the face of common hazards like earthquakes, mudslides, and floods (the threat of hurricanes and tsunamis was not particularly high that far inland, though they were more of an issue on the state's coast).

While officials told me that Guerrero had one of the more respected Civil Protection institutions in Mexico, this power was not immediately visible in its offices. The ceiling of Guerrero's office of Civil Protection in Chilpancingo drizzled an inconsistent spatter of dirty water that had, when the building's drains were clogged in a storm, simply collected on the roof and was slowly filtering down upon us. Like Oaxaca, Guerrero is among the most seismically active and poorest states in the nation.[58] It was the site of the first field stations in the Mexican earthquake early warning system, and after Mexico City and Oaxaca, it was the next to start disseminating earthquake early warnings. It also suffers storms, mudslides, tsunamis, floods, and dangers from incendiary materials like the gas canisters people use for cooking.

The Civil Protection offices were busy with people working to improve the structural components of buildings, enforce building regulations, support governments' abilities to maintain crucial services in the event of a crisis, and change individual priorities and social behavior. There, my questions about earthquakes and earthquake risk mitigation were put in context of broader projects of risk mitigation focused explicitly on social life. "We cannot predict quakes," one thoughtful official in Oaxaca calmly explained, "but our job is to build a culture."[59] He and his colleagues were tasked with more than earthquake risk mitigation. They explained "culture" as a key concept they used in their work to make Mexicans safer.

There are many ways to define "culture." As anthropologist Michael M. J. Fischer explains, the concept has experienced "historically layered growth of specifications and differentiations . . . that . . . allow new realities to be seen and engaged as its own parameters are changed."[60] For anthropologists like myself, "culture" often suggests the habits of life and mind that people receive, rework, and reproduce, as well as the structures and forces that frame how they do so. The term is not anthropology's alone to work with, however, and it has other meanings outside of our research.[61] In the offices of Civil Protection, "culture" means something very specific. Some of their materials defined it as "the set of knowledges, beliefs, abilities, attitudes, and values related to earthquakes which, in a given place and moment, every community has as a product of its historical experience."[62] Civil Protection's founding documents described "culture" as an optimal site where an intervention can happen to both reinforce aptitudes that already exist and teach new abilities, while elevating "aspiration and creativity."[63] By 2012, Civil Protection authorities had been tasked not only with recognizing and fostering communities' existing capabilities or strengths but also with developing new and useful elements of culture by educating people and encouraging them to change how they live.[64]

Many Civil Protection officials and commenters in the popular media thought they had a long way to go to reach such goals. They described a public uninformed about and unprepared for hazards and as "uncultured." In conversations and systematic searches of news archives, I learned that many people thought Mexicans had no seismic culture (or risk culture, self-protection culture, or prevention culture) at all.[65] In fact, rather than describing culture as certain sets of knowledge, beliefs, and attitudes that might not include as much engagement with risk mitigation strategies, these accounts suggested that Mexico simply did not have culture relevant to risk mitigation at all. I asked Civil Protection officials about it when I met with them. In a community with an adequate "culture," I learned, ordinary people understand their roles and responsibilities, and incorporate strategies for risk mitigation into their lives. People with adequate cultural awareness would know how earthquakes occur and where they are likely to be felt, and would respond appropriately to that knowledge. Taking part in drills, knowing and implementing general strategies to make buildings safer, and developing emergency plans and the ability to stay calm in an emergency were all key activities that Civil Protection officials associated with developing a culture of preparedness. All the work they did to cultivate such a culture, however, never seemed to have satisfactory results. By and large, Civil Protection officials told me, most Mexicans were not aware of or committed to seismic safety—or, indeed, any form of risk mitigation—in this way. Amid discussion of an absent seismic culture, there is, of course, an occasional reference to such a culture that may once have existed in a pre-Hispanic Mexico. Inevitably, these references to a mythologized and unified historical past rarely include much detail.[66] I have asked experts in disaster prevention if positive examples exist in Mexico. While perhaps they do, I have yet to learn of any.

Many times, culture is depicted in deficit, with outreach envisioned as a one-way communication process in which experts simply distribute information and try to instill correct priorities and elicit certain behaviors. Science education and communication researchers have critiqued deficit models like this. People are not passive recipients of knowledge, and treating them as such without substantially engaging the variety of ways in which people might encounter the world not only privileges elite experts' perspectives over those of others whose lived experience—and indeed, lives—may be at stake, but it is also an ineffective strategy for communication and cultural change.[67]

There are further implications, though. Addressing culture as a deficit means placing responsibility for acting to mitigate unsafe conditions on ordinary people. When these conditions are the product of broader physical and social issues, such as poverty and marginalization, then individuals have limited opportunities to make significant changes.[68] As critics of such strategies point out, "scapegoating" people for how they are exposed to danger means treating those who suffer harm as passive, indebted, or simply bad victims for pursuing their own goals and livelihoods.[69] By this logic, vulnerability to earthquakes becomes the result of failures of the Mexican people, not the state, which creates and enforces safety regulations. Mexican disaster scholar Jesús Manuel Macías has critiqued Civil Protection institutions on precisely this issue, writing that they "transfer responsibility for the protection of life and property from a state authority to the population at risk."[70] Some Civil Protection officials, however, including those I met in the dripping and dirty office in Guerrero, spoke passionately about what they wanted to do for people and about their careful efforts to engage communities in appropriate risk mitigation practices.

What a tangle. In a Starbucks on busy Insurgentes Street in Mexico City, I laid all these issues out before Alejandro Martínez, an engineer and Civil Protection official who had been involved in seismic risk mitigation work for many years. He was generous with his insights and easy to talk to, so I described my confusion. What was I supposed to make of how some people approached risk mitigation as if ordinary people were always failing at keeping themselves safe? How could I understand approaches to reducing vulnerability through culture that some saw as engaging with community needs and practices and others used to scapegoat the very people most at risk? After traveling to Oaxaca and Chilpancingo, sitting down for meeting after meeting, and collecting and evaluating articles from local newspapers, I knew this was a key issue. These ideas about Mexicans and their culture formed the backdrop for the development and growth of SASMEX.

Martínez was sympathetic. He explained the use of the "culture" concept that he observed as a sort of heuristic, a way of drawing attention to issues that could be addressed in the context of incredibly limited resources and funding. "*Our* culture," he told me, using the term now to describe the beliefs and practices of those who worked within Civil Protection, "is to identify problems and work for the future."[71] It is true that Mexico suffers from corruption and a limited ability to enforce building codes, and that many Mexicans do not have the resources to build and maintain safe spaces. An earthquake early warning can be used to address this conundrum.

An earthquake early warning system makes sense as one of many efforts to reduce vulnerability and is particularly revealing as an effort to put responsibility for personal risk mitigation on ordinary people in the absence of other resources. Mexico is a poor nation in comparison to others represented at the International Conference on Earthquake Early Warning, and others have different logics and institutions of risk management into which such a technology fits. Taken on its own, an earthquake early warning system may put the onus on ordinary people for preserving their own safety when there is a limited amount that they can do if the agencies and organizations around them are not also working to reduce their vulnerability and support resilience by hardening infrastructure, enforcing regulations, or providing financial support to struggling people.

THE TECHNOLOGY EXISTS

At the third International Conference on Earthquake Early Warning, the then California state senator and earthquake early warning advocate Alex Padilla commented, "I don't want to be here after the next big one wondering why we didn't implement a system when the technology existed."[72] Certainly, tools for earthquake early warning exist, but the implication that they can simply be implemented in one place as they have been in another is misleading. Although technology advocates do not always take the time to unpack exactly why they believe a given tool makes sense, technology development relies on certain ideas that are both shared internationally and, as I have shown, are specific to a nation or community.

Padilla's comments were like many discussions of this technology at the conference in that they described earthquake early warning systems as largely technical. In the worst articulation of this popular logic, people developing effective networked sensors and algorithms for early warning need not concern themselves with considerations of user perspectives and needs. Certainly, the diverse technical innovations that make earthquake early warning possible are worth celebrating and mobilizing to inspire new implementations. Treating risk mitigation technology in this way, however, while also explicitly considering reduction of human vulnerability as its core purpose, creates a difficult, almost paradoxical, situation. Here, people are the central beneficiaries of these technologies.

Earthquake early warning technologies makes sense to so many as tools to change how people at risk live in hazardous environments. Paying attention to how people may be vulnerable to hazards is essential to modern perspectives on risk mitigation around the world. Just as vulnerability can mean both physical exposure and poverty, however, addressing it can mean very different things: for example, giving ordinary people or governments responsibility for risk mitigation, or engaging with communities' particular practices and needs, or denigrating them as deficient by definition. These concepts inform technology development and adoption. Nonetheless, those technoscientists concerned with making changes in how people experience earthquakes may share only the loosest understanding of what a risk mitigation technology can do.

4   INTEGRATING INFRASTRUCTURES

I learned what trouble the infrastructure for early warning could be in an ordinary meeting of high-level administrators at CIRES headquarters. The popular smartphone app SkyAlert issued an earthquake early warning out of the blue. In the meeting room, smartphones around the room buzzed as people texted to let us know that they had received warnings. Elsewhere, thousands of people in Mexico City rushed out of buildingss and into the streets.[1] The speakers mounted on the walls around us remained silent, though. We were all usually prepared to take an earthquake early warning seriously, but this one did not add up. If there had truly been an earthquake sweeping inland from the west coast of Mexico that day, detected and analyzed by SASMEX stations, the building would have resounded with *¡alerta sísmica!*

Since we did not hear the distinctive warble, we remained seated, certain that we were in no immediate danger. No matter what the apps on our phones said, or what text messages we received, we were confident that no earthquake had been detected ripping out of the subduction zone to the west or the faults deep under the mountains to the south. After all, at that time, SASMEX, administered from the very building we were sitting in, had the only network of seismic stations that could provide early warnings to Mexico City.[2] There was nowhere else a real warning could come from, if it was not coming from CIRES instruments.

As it happened, we were right to be skeptical. As seconds passed, the silty, seismically sensitive soil below us remained still. Soon enough, we could put words around what had happened: a misfire. At 12:16:54 p.m. on July 28, 2014, a popular smartphone app called SkyAlert had issued a false alert. If there had truly been an earthquake that day, the SkyAlert app's warnings would have joined a chorus of alarms echoing across Mexico City. SASMEX's broadcast would have triggered howling sirens throughout CIRES headquarters and in hundreds of other buildings around Mexico City and other user communities. While there were not yet sirens on street corners in 2014, radio and television stations would have blared warnings, too. If the earthquake in question had indeed whipped out from the Pacific coast, where the false alert suggested, we would have had around one minute between when the sirens began blaring and when the comparatively slow seismic waves began to reach us, plenty of time to get to a safer place than the second floor of a converted residential building.[3]

Although none of us at CIRES evacuated in response to the SkyAlert warning, the meeting room I was sitting in soon emptied. The two dozen engineers, technicians, and administrators present for the meeting abandoned their weekly agenda. They went to work, instead, on managing the effects of the misfire event. Mexico City's population was living not only with earthquakes but also with multiple forms of earthquake early warning, and SkyAlert reached over a million people who had subscribed to earthquake early warnings via its app.

Later, company representatives at SkyAlert helped me understand what had been happening at their headquarters when the false alert was triggered. The team showed me the signal they had received from CIRES on July 28. The message in question had appeared to the SkyAlert system as all signals from CIRES do: as a series of alphanumeric characters indicating the date, time, and nature of the message.[4] They showed me on a computer screen how the series of characters in the test message, the kind that they and the other organizations receiving CIRES alerts multiple times each day, had been garbled—letters in the alphabet replaced by @ signs to show an absence of information (see figure 4.1). The SkyAlert system extrapolated from what it had, though, and it found a pattern. This particular message had ordered its code for date and time in such a way that, with the changeable interference of city life, SkyAlert's system was able to register the message it received as an earthquake early warning. This is the genesis of the SkyAlert false alert: simple bad luck. There was no protocol for SkyAlert to confirm receipt of a message from CIRES or make sure that their system had interpreted it correctly, so the app automatically pushed a warning to client smartphones. In

FIGURE 4.1
A record of CIRES signals interpreted by SkyAlert's system. *Source*: Author (2014).

a complicated situation, they explained, a simple technical error and a lack of coordinated confirmation procedures made the misfire.

Such a neat technical explanation belies more complex conditions of possibility. I heard few arguments that that this kind of event was simply an unfortunate part of life with earthquakes and earthquake early warning systems. Instead, many experts in the disaster risk mitigation community, authorities, and Mexico City residents treated the false alert as far more than a technical glitch to be marked, analyzed, and repaired. It was a serious problem, disturbing to the technical experts involved. They were deeply concerned that people who lost faith in earthquake early warnings would not respond effectively to hearing them in the future. Some told me that early warnings were designed to save lives, and that could only happen if people were willing to trust them. Events like this were worrisome indeed.

Advocates explained their worries about false alerts in terms of a "crying wolf" effect.[5] In the weeks after the incident, people at CIRES and elsewhere told me that the misfire could not only jeopardize trust in earthquake early warning, and consequently impede response to subsequent alerts about real threats, but it could also have consequences for any project of earthquake early warning in Mexico. They worried that the misfire would undermine any effort related to earthquake early warning—whether operated by Sky-Alert, CIRES, or other agencies. The maintenance of the earthquake early warning system required ongoing investment, whether in the form of app purchases or contracts with the governing bodies of user communities. Additionally, this was not the first time that an early warning app had misfired. It was not even the first false alert propagated by an app while I lived in Mexico City. The frequency of hiccups in the integration of different systems was alarming to the engineers responsible for making early earthquake warning work.[6]

In this chapter, I show how engineers shoulder the burden of making what many assume to be integrated infrastructure functions. The SkyAlert misfire stands as an indicator of just how challenging it is to maintain multiple communication channels in the context of a semi-articulated system. Common principles in emergency communications suggest that sending a warning through multiple channels is an important technique to reach different publics. These principles, however, often disregard the context of the system that actually propagates the warnings. Here, diverse channels of communication come along with confounding conditions. Technology development at CIRES and SkyAlert were siloed and structured by competing priorities about public risk management, and this created conditions for incompatibilities. By describing the misfire in 2014 and the responses of technoscientific professionals at CIRES and SkyAlert, I demonstrate how semi-integrated infrastructures form the basis of conflict. I then turn to a rumor of sabotage that I encountered shortly after the misfire. While I found no evidence that this rumor was true, its mere existence is important to note. The rumor's elements shed light on the misfire and, more broadly, on what happens when channels for warning communication multiply. This single controversial event reflects many challenges related to developing a system to support new ways of engaging with earthquakes.

A PARADOX OF PROLIFERATING CHANNELS

The SkyAlert app was different from anything offered by CIRES in 2014. While CIRES, a state-funded NGO, broadcast general alerts using radio signals to companies, schools, and TV and radio stations, SkyAlert was a for-profit company and pushed personalized messages to the smartphones of subscribers through wireless data connections. Although both were run primarily by engineers, they could reach people in different situations and appealed to different populations. These differences are, theoretically at least, a good thing; they could reach a broader audience together than either could alone, and if people received both, the warnings could reinforce messaging.

Emergency communication scholars suggest that good alert messages should generally be consistent, clear, and offer guidance. It should, moreover, come from sources considered credible by anyone who is meant to use information provided.[7] When populations receiving emergency messaging are diverse, content must be suitable for many different needs. It should come to people through appropriate means, ideally from multiple sources, and should be followed up with communication that explains the event in more detail.[8]

Engaging and supporting multiple modes of alerting has the potential to create tension, though. This tension has marked the development of the semi-integrated infrastructural system in which CIRES, SkyAlert, and other agencies develop and distribute early earthquake alerts to Mexican people. In practice, the distinction between a consistent message that has been modified for diverse people and an inconsistent message is muddy. Some might even see the same earthquake event described in different ways—perhaps in intensity by one channel and magnitude by another, or projected to reach different magnitudes at their source or intensities as experienced—and consequently be confused.

For alerting authorities, the potential for others to adopt warning tools that do something significantly different from their established practice is real and worrisome. This issue is not limited to Mexico. In California, for example, Los Angeles mayor Eric Garcetti's choice to lower the threshold for ShakeAlert earthquake early warnings was met with consternation from emergency managers who wanted communication to be clear and consistent throughout their system.[9] Contests over messaging become further complicated when private technology vendors get involved. When warning channels proliferate, some providers may seek to expand their user base by undermining others' work in a competition for limited attention and money.[10]

While warning channels may compete, many people do not have significant choices about what system they use. Earthquake early warning systems are rarely either technologically or conceptually straightforward. In Mexico, most people have not historically received warnings directly from SASMEX. For many years, warnings went out primarily via television and radio, making access dependent on whether people were listening to broadcast media at the moment an earthquake began. Apps do the same work as intermediaries between CIRES and people at risk of earthquakes. Smartphone use is high in Mexico and growing fast.[11] Around one-quarter of people in the nation (or more, depending on use practices) had access to smartphones in 2014, when the false alert happened. More recent reports put that number at approximately two-thirds.[12] This trend has offered a new point of access to a warning signal already broadcast publicly in the seconds, or sometimes minutes, before earthquakes strike some major metropolitan areas. Because relatively few Mexicans living in these cities had access to the specialized radio receivers located in schools, hospitals, and television studios, apps like SkyAlert could fill an important gap and have plenty of potential customers.

Of various smartphone apps doing earthquake early warning, SkyAlert was among the most successful. As of 2014, the app had over a million paid subscribers and a substantial social media presence. For $3.99 USD, users could interact with the clean, active interface on their smartphone screens to learn about earthquakes and extreme weather events. They might also receive recent news items and shortcuts to Mexican emergency services numbers, elegant against a geometric orange and black branded background. When the misfire happened, SkyAlert was still only just beginning to build the utilities that the app offers today.[13] At every opportunity, though, SkyAlert's developers had already found ways to differentiate their project from that of CIRES.

SILOED DEVELOPMENT OF WARNING TECHNOLOGIES

SkyAlert began as a project of SkyTel, a prominent name in the Mexican telecommunications industry since the late 1980s. SkyAlert's founder, Alejandro Cantú, the son of the SkyTel CEO, had trained as an industrial engineer at prestigious Iberoamericana University. The family company's reputation and financial support bankrolled the new project's development. Only two years before I began my fieldwork, outside investors had helped SkyAlert became a fully independent company with ambitious plans to take processing to the cloud, build a seismic sensor network independent of CIRES, and allow users to choose the warnings that they would receive. The SkyAlert app was only one of SkyAlert's products, but it was core to the company's vision for transforming emergency communications in Mexico.

When SkyAlert came on the scene, CIRES had been operating an earthquake warning system entailing robust accelerography, data analysis, telecommunications, and broadcast to communities across Mexico for more than two decades. SkyAlert launched its app as a tool that received messages broadcast by CIRES, supplemented with data from its own small but growing seismic network, and pushed warnings to users. However, the app was designed less to build out the SASMEX system than to make something wholly new. The SkyAlert team may have identified an opportunity in relation to what CIRES was doing, but they developed their technology largely independently of CIRES. This allowed SkyAlert and Cantú, now the CEO of the new company, to build an app that relied on very different logics than the public earthquake early warning system did.

The priorities that informed the SkyAlert app's development were sharply distinct from those of the SASMEX system it interfaced with. One basic conflict stemmed from the different ways that SkyAlert and CIRES understood the people who might benefit from warnings. SkyAlert considered users to be individual people. SkyAlert's app offered users ways of being intimately and immediately engaged with earthquake warnings. Users could, for instance, choose to test to see if their device was connected to the SkyAlert system with the press of a button. The smartphone format offered users greater flexibility and more information, allowing them to customize their messaging and learn more about an earthquake. This was markedly different from the CIRES mode of alerting: a siren howl with little detailed information, broadcast to anyone and everyone at once.

I learned about these differences at CIRES and my understanding of their importance grew more nuanced when I visited SkyAlert's offices in Iztacalco, on the east side of Mexico City. SkyAlert's space was smaller than CIRES's, had accumulated less equipment, and boasted fewer people moving about. I went to tour the office on a rainy day in August, just a month after the misfire. I heard about novel plans under way to develop new earthquake early warning functionalities, a sharp contrast to CIRES's efforts to simplify and standardize earthquake communication. The messages that SASMEX broadcasts are coded to indicate where a quake originated, whether an earthquake detected is worth alerting, and, if so, whether it is likely to be of moderate or strong intensity. Its sirens, however, can do one of two things: go off or stay silent. Cantú himself explained how different his company's approach was from the public earthquake early warning system simply, saying, "It's not about me *deciding*. We're providing information, we're providing them with things they want to know."[14] His goal was to offer options and information, not just alerts.

Taking a coffee break, Cantú introduced me to a few engineers on his team. They were full of enthusiasm for the project. The app, one explained, was a self-conscious exercise in Mexican innovation. Another described it as revolutionary. Then-president Enrique Peña Nieto's government was already actively reorienting the Mexican state agenda around digital innovation, and these engineers told me that they saw SkyAlert as part of a national movement.[15] This point of contrast was as meaningful from CIRES's perspective as it was from SkyAlert's. CIRES had institutionalized processes of incremental system development, and insightful engineer Antonio Duran even explained to me, very seriously, that "if it is obsolete, then it works," was one of the key principles or, in his words, "laws," of CIRES.[16] New technology might be unreliable, but CIRES engineers and technicians could be confident in older technology and in their ability to make repairs if something did go wrong. This axiom was not conducive to the kinds of digital innovations that excited Cantú and his team. Where SkyAlert was self-consciously innovative, CIRES relied on old, well-vetted technologies and procedures. Where SkyAlert pushed notifications to individuals, CIRES broadcast to populations. Consequently, it was no surprise to me that the organizations' products were similar but distinctive. Nor was it a surprise that they only coordinated their work when necessary. The misfire, however, showed the trouble that could occur at their interface.

Scholars of infrastructure have long insisted that no matter how concrete or organized they seem, we must remember that systems are changeable and contingent. Sociologist Susan Leigh Star famously argued that while infrastructures like the system by which CIRES signals were broadcast to SkyAlert certainly appear easy to ignore until they break down, there are people who never assume a system will work: those responsible for its maintenance.[17] The challenge of integrating CIRES's signals with smartphone apps complicates the idea that infrastructure is either working and invisible, or broken and salient. Certainly, the misfire was an emergency that commanded international attention, moving the technoscientists working at CIRES and SkyAlert to take quick action. The breakdown forced CIRES and SkyAlert engineers to consider aspects of their respective systems that they typically took for granted, and further required that they explain them to publics who had very little sense (if any) of how earthquake early warning worked and how the various warning channels were related. And yet, it would be a mistake to say that earthquake early warning infrastructure worked smoothly before the misfire; indeed, integrating multiple channels of warning delivery makes the very idea of "working" seamlessly and smoothly an elusive if not an impossible goal. Instead, in a complex system of related networks, the practices and priorities that connect an organization like SkyAlert to one like CIRES are constantly on the verge of potential breakdown as they react to internal and external conditions.[18]

Anthropologist Brian Larkin has defined infrastructures as "things and also the relation between things"—a broad definition to say the least.[19] Nonetheless, this definition is a particularly useful way of thinking about how infrastructures work. It suggests that the system that broke down on July 28, 2014, is both a topic for inquiry in and of itself and an incident through which to consider the shifting, changeable relationships between different early warning systems. Proliferating channels for disseminating warnings offer the promise of supporting emergency communication, but without means to regulate how they interact with one another, their existence may come at the expense of the mutual trust and coherent processes that a flexible and integrated warning system requires.

RESPONDING TO A MISFIRE

How can channels for emergency communication that are so different and so incompletely integrated co-exist, each asking people to trust them to signal oncoming earthquakes? After the SkyAlert misfire, CIRES and SkyAlert had to negotiate their relationship, and determine what publics they sought to serve—including those who were already making #NoEraSismo

(#ThatWasNoQuake) trend on Twitter and talking about how things may have gone wrong.

As soon it became evident that the SkyAlert app had sounded a warning for a quake that was not coming, the CIRES team leapt into action. They began by investigating the extent of the misfire. They picked up their phones, showing each other tweets and messages they had received. Some began making calls to policy makers, emergency managers, and representatives with other organizations focused on public safety around the city. News of the false alarm was circulating widely.

The meeting table emptied as people broke off into various conversations and errands. CIRES had never encouraged the companies that now promised to disseminate their early warnings via apps; most of their engineers regarded smartphone notifications with skepticism, at best. But how had the misfire happened? When had the device that SkyAlert used to receive and process coded CIRES signals last been serviced? People checked their records and debated how to respond. The usual scene at CIRES offices included dozens of people, some sitting at their desks and others moving between departments, talking to their peers and collaborators; but now everyone seemed to be moving at once, and with purpose. Some continued with ordinary projects, certainly, but many set aside the tasks that they had planned for the day. Even though the erroneous message had come from SkyAlert, SASMEX was not in the clear. "They're going to say it was our equipment that failed," one man told his colleagues, concerned about the CIRES system's reputation.

At 12:51 p.m., less than an hour after the misfire, SkyAlert's representatives publicly declared that the app had responded automatically to a message from CIRES's system. Their tweet was soon quoted in many popular news outlets: "Our platform received a seismic activation by CIRES, we are investigating the causes."[20] Several of the senior CIRES administrators went to the Public Outreach Department to work on a response. This department was based in a small room ringed with desks and computers. The CIRES leaders crowded around the software engineer who managed CIRES's social media accounts. Today, the stakes of the organization's online presence were higher than usual. Together, the senior administrators composed careful messages, dictating to their junior colleague as she typed and posted them on social media to deny any culpability.

CIRES needed to distance itself from SkyAlert to make people understand that the misfire was not the organization's responsibility nor an opportunity to indict earthquake early warning in general. The CIRES team was concerned that any fault the public attributed to CIRES would have implications for the entire system. The organization could lose credibility with people who should respond to warnings, and CIRES might lose contracts with the governments that enabled its work. CIRES, a nonprofit, funded the maintenance and expansion of the entire earthquake early warning system through annual contracts with user communities. These contracts meant the system's on-going viability was never certain and was subject to the changing sentiments of policy makers. If CIRES lost contracts, then the municipalities of Mexico City, Oaxaca, Chilpancingo, Acapulco, Puebla, or Morelia might stop broadcasting earthquake early warnings to their residents through SASMEX. They mounted a spirited social media response to defend the system's reliability.[21] To do so, they not only addressed the misfire, but also called the SkyAlert app's ordinary function into question. I watched their messages post live from a screen a few feet away.

"The #SkyAlert #SkyAlertApp is not well-integrated," the CIRES accounts on Twitter and Facebook proclaimed. "Lags have already been identified. They do not disseminate the alert simultaneously to their subscribers."[22] The wording that they had chosen referred to both SkyAlert's ordinary processes and priorities as well as the misfire itself. This issue—"lags" as opposed to "simultaneity"—invoked a long-standing topic of some concern to CIRES's engineers. Lags are moments of jarring disjuncture between various temporal operations.[23] SkyAlert's data indicated that the false alert reached less than half of the 1.5 million smartphones with the app in the first five seconds, and had still not reached all of them after 15 seconds.[24] One subscriber's phone might have jangled with an alert in time to take shelter, while her neighbor might not have received the message until after the time of the predicted earthquake had passed.[25] For the technical experts at CIRES who understood reliability to be key to effective early warning, SkyAlert's lags alone made the product unreliable and therefore a reason to distance themselves.

Lags, like false alerts, presented more than a reputational problem, CIRES engineers explained. Lags could generate confusion. People who receive delayed messages might expect a new earthquake event that would never arrive, thereby reducing trust and future willingness to respond. Lags were antithetical to the kind of early warning that the CIRES team wanted to provide, but the proliferation of smartphones had facilitated new forms of interaction with earthquake early warnings that CIRES could not easily regulate. The integration with smartphones destabilized the principles of early warning that SASMEX was built around. Smartphone apps could give warnings to people who were far from radios, televisions, or dedicated emergency signal receivers,[26] but they were unreliable. While the lag was a considerable concern, there were more simple ways smartphones might fail to warn people—if they ran out of batteries, were placed far away, if in-app alarms were somehow silenced by other phone settings, or if they simply had patchy data connections.[27] Even so, these potential pitfalls did not dim SkyAlert representatives' enthusiasm for disseminating warnings through smartphones.

SkyAlert's leadership did, however, recognize misfires as a concern. On my visit to the business's headquarters in the autumn of 2014, I heard that the misfire was bad for business and public safety. If people lost confidence in SkyAlert's app service, they could stop subscribing. While SkyAlert was working with government organizations, they were not as closely involved in a public, state-supported project as their CIRES counterparts were. Sky-Alert and CIRES representatives all spoke with me, as they did in other public messaging, about civic responsibility. Both groups, like the people at the conference with which chapter 3 opens, referred to the opportunity to save lives when they discussed their projects' promises.[28]

For all that they used a similar language that referred to high stakes, SkyAlert engineers measured their success differently than CIRES engineers did. Cantú told me that whatever the lag in SkyAlert's system, it was worth it. He thought that SkyAlert was crucial part of Mexico's earthquake early warning infrastructure. He explained that his system had recorded 92 percent of all users receiving messages and opening their apps within fifteen seconds of the misfire. If nothing else, he argued, this had been an opportunity to see how the system worked. He did the math for me. If they had 1.5 million users, a 92 percent response rate would mean alerting more than 1.38 million people. That was excellent, to his mind. "I'd rather alert 1.3 million people than three hundred thousand," he said, comparing the fraction of SkyAlert users who got the false alert to a much lower number that he imagined the CIRES broadcast could reach without the help of apps like his. Rather than trying to create a lag-free experience, Cantú argued that it was his obligation simply to put earthquake early warning into as many hands as possible.[29]

CIRES and SkyAlert's responses to the misfire demonstrate how siloed system development can provide the basis for divergent understandings of earthquake risk mitigation. As it happened, Mexico City government authorities supported CIRES and called for sanctions of SkyAlert,[30] but both organizations were able to continue their work. The question of managing the margins of the systems, where different alerting priorities and system processes made miscommunications possible, remained neglected. While people I interviewed at SkyAlert and CIRES focused on the conceptual and technical distinctions between the two systems that they had observed, other commenters hypothesized about actions that these differences might inspire.

MAKING SENSE OF MISFIRE

After recriminations from Mexico City's leadership, SkyAlert soon formally and publicly accepted culpability for the misfire. The company's official account tweeted "We offer sincere apologies for this unfortunate incident."[31] For some, the issue was still unsettled, though. A few people professionally invested in earthquake risk mitigation continued to discuss the incident and its effects. The explanation advanced earlier in this chapter, which presented the false alert as the product of bad luck, was insufficient for them. These people continued to think through the case in light of what they knew of CIRES, SkyAlert, the relationship between the two organizations, and Mexico.

Some suggested that the misfire had not been a technical malfunction at all, but rather a strategic act of sabotage on the part of the CIRES team to discredit SkyAlert. I was surprised to hear technoscientists who were active and respected in the earthquake early warning community discussing this theory. However, several spoke with me about this in serious tones. "I think that this could have been a huge coincidence," one man admitted; but, he continued, "It could have been something very evil."[32]

This man, who I will not describe further here, put his concerns plainly. He was, however, not alone in suggesting that CIRES might have intentionally caused the 2014 misfire to discredit SkyAlert. There were others like him—professionals who were skeptical of how SASMEX operated. CIRES engineers, their theory went, might have intentionally composed and sent the message that caused the SkyAlert system to issue its false alert. For those who voiced this conspiracy theory, the misfire itself was a sign of human action. They understood it as an effort on the part of the CIRES team to manage the infrastructures of earthquake early warning; to control who could participate and how. If SkyAlert lost the precious trust of users, then the company might shift its focus away from its app and stop generating this kind of push-notice alert. In that case, CIRES would have fewer rivals challenging their model of alerting and reduce the semi-integrated systems troubling their efforts to keep early warnings trustworthy and simultaneous.[33]

When I first encountered the theory that the 2014 misfire might have been sabotage, I struggled to make sense of it in the context of everything I knew about the CIRES team's worries about muddled emergency messages. First, sabotage would be logistically difficult to pull off reliably—was I to understand that these saboteurs imagined a potentially troublesome coded test message, and then waited for both the date and time for it to be pertinent and the right unpredictable circumstances of interference to confuse things, and then when it did, acted out an emergency? It seemed unlikely that anyone at CIRES could pull all that off, even if only a few people were in on the conspiracy. Few actions at CIRES were taken without the approval of the NGO's leadership, and the leaders I saw during the misfire event did not look to me like people whose elaborate scheme had come to fruition. Beyond that, though, I sincerely doubted that the people I had met at CIRES would ever consider it possible to damage SkyAlert's reputation without harming that of earthquake early warning itself. Many people at CIRES had spoken with me at length about how few members of the public could really understand SASMEX or distinguish between CIRES and SkyAlert. My ethnographic research corroborated their sense of popular earthquake early warning awareness: of the Mexican residents I have spoken to since 2011 who have no connection with seismic risk management, only a handful have been able to clearly articulate the distinctions between the systems. Members of the CIRES team had also voiced deep concerns about "crying wolf" effects to me during my time with them. Their concerns that people could lose faith in early warning, combined with their awareness of just how much CIRES and SkyAlert were elided in popular imagination, made me very skeptical of the conspiracy theory.

The more mundane explanation, which I offer at the begining of this chapter, takes the incident as an utterly ordinary product of under-integrated system lacking confirmation procedures. It was insufficient for some. In the weeks and even years since, I have been called upon to tell the story of the interrupted CIRES meeting multiple times. People involved in disaster risk mitigation have been especially interested in hearing how leading figures at CIRES reacted to the news. My observations about their surprise, the distinctions between technological goals of CIRES and SkyAlert, and jagged interfaces between the two systems have not always persuaded skeptics of the CIRES team's innocence. For all that my research demonstrated the limitations and inconsistencies of Mexican state support for CIRES's earthquake early warning system, critics still considered CIRES "insiders" with special connections and influence over matters of public safety. To them, an act of intentional sabotage represented just one more instance of corrupt Mexican elite technoscientists allied with the state trying to maintain their privileged status, exclude outsiders, and undermine approaches to alerting that differed from their own.

Although I find the sabotage story unbelievable in and of itself, the fact of its existence is important to note. Most crucial to this case, when people engage in speculation about corruption, they showcase underlying ideas and attitudes.[34] This theory and its persistence should not only be considered in relation to empirical events—that is, as a potentially factual or false claim—but rather, it should be understood to signal efforts to think about the necessarily social nature of technology.

When people suggest that CIRES engineers sabotaged SkyAlert, they do so in the context of a long and well-documented history of the power that elite technical experts have in Mexico.[35] They can obtain political favor, which in turn can mean influence over policy decisions large and small. Some political maneuvers are obvious to astute Mexican publics, and others are less so. Nonetheless, many simply assume that such activities are happening even in the absence of evidence.[36] They are the topic of rumor, or *chisme*. Anthropologist Claudio Lomnitz has described *chisme* as a key part of "alternative communicative relationships" as necessary to Mexican political life, crucial for navigating a complex and secretive world.[37] When playing upon political relationships is considered crucial to maintaining the support of policy makers, it is little wonder that the sabotage theory found such traction.

Talking conspiracy like this means putting social relationships at the center of conversations that are otherwise focused on technologies alone—and as I have shown elsewhere in this book, attempts to bracket off technologies from the social and environmental world lead to limited and unsatisfactory analysis. The motivation that conspiracy theorists who spoke to me offered to explain the sabotage they proposed was the CIRES's interest in controlling the systems involved with earthquake early warning. The theory, however dubious, addresses the complexity of operating an earthquake early warning system. Its very existence highlights the frustrations of infrastructure integration that often fall out of more narrow accounts of risk mitigation technology.

SILOED EFFORTS CONTINUE

Well-established principles of emergency communication suggest that it's better to have more ways to get an emergency warning—more "channels"—than it is to have fewer. A warning system benefits from multiple ways to reach people. More channels means reaching more people or reinforcing a message's power by making sure that a person can encounter it multiple times, in different ways. In practice, however, the divergences between channels, both technical and conceptual, create new problems. SkyAlert is a different channel than a television broadcast, so theoretically it should be valuable. However, the engineers at CIRES worried that SkyAlert's negative impact on public trust for earthquake early warning would outweigh the benefit of having an additional channel. Without adequate public education efforts or coordinated messaging, multiple communication channels seemed dangerous for warning efficacy rather than advantageous.

Technologies, and infrastructures, change. Mexico's public system started to broadcast through radio, then began setting off sirens. Integrating mobile technologies has become increasingly important as use proliferates, and new technologies mean new partnerships. Regardless of the tools or partners involved, systems of emergency communication are always emergent. Like any other form of infrastructure, their state is in flux. They may break down or misfire, potentially producing confusion. Reflecting on these incidents allows us to consider how technological systems can be both deeply related and siloed.

At CIRES headquarters and outside its walls, people told me that they were concerned that in a real earthquake they would be slow to take cover or evacuate if they doubted the veracity of an alert. On Twitter and Facebook, commenters also focused on shorter-term consequences of the incident for their technologically mediated relations to the seismic environment. People berated SkyAlert and earthquake early warning technologies in general for scaring them; for creating a panic that might itself be dangerous to their health and well-being. These comments may have been deployed hyperbolically, but they did display concern regarding the effects that a false alarm, and an earthquake that had not happened, could still have.

In these conversations, commenters put the onus of successful warning almost entirely on technical experts and their separate, siloed efforts. This was an impossible task in the face of inevitable breakdowns. Perhaps technoscientists *could* integrate their technologies and work together to support new ways for people to live with earthquakes, but the misfire and events that followed it demonstrated just how radical a change in approach to early warning would have to be.

Since 2014, SASMEX's infrastructure has grown to include public sirens. SkyAlert has limited its involvement with the public system, developing a private network of 120 low-cost sensors to generate information about oncoming earthquakes rather than relying on CIRES's more sophisticated equipment and its public broadcasts.[38] While SkyAlert's efforts promote earthquake early warning broadly, they are now even less integrated into a coherent infrastructure with CIRES than they were when I was doing fieldwork. While this reduces the chance of a misfire of the sort that happened in 2014, it also reduces the opportunity for any sort of consistency in alerting publics that may be confused by the two now-parallel systems, which promote warnings about the same earthquakes in very different ways. Recent earthquakes and ongoing seismic activity in Mexico have made further growth of both CIRES and SkyAlert's earthquake early warning infrastructure possible. Between them, more people have access to emergency messaging. However, no matter how these infrastructures change, siloed efforts may mean that similar challenges will remain.

III Engineering Risk Mitigation

5   MEASURING EARTHQUAKES LIKE ENGINEERS

The technoscientists whose work I describe in this book share broad understandings of Mexican seismic phenomena. Their descriptions of where seismicity originates, why Mexico City is so sensitive to it, and how earthquakes endanger human beings are generally very similar. Some people with whom I worked most closely, however, took care to articulate significant distinctions between different technoscientific approaches to intervening in life with earthquakes. Dr. Edmundo López, a senior member of the CIRES community who was responsible for many day-to-day operations, made this very clear when I asked him to explain the system. The important thing, he said, pushing his big glasses up his nose, was that engineers had developed Mexico's public earthquake early warning system, not scientists.

López did not constrain himself to communicating with words. Instead, he pressed his hands to the wall of a CIRES office theatrically to show me what he meant. "Look," he said, "I'm measuring like an engineer now."[1] He walked his hands along the wall, lining them up thumb to thumb, little finger to little finger, until he had moved along the whole length of the wall between an overcrowded pasteboard bookshelf and the copier, miming measurement with his hands alone. "And now," he said, "I'm measuring like a scientist."[2] On the word "scientist," he changed his whole bodily orientation toward the space, getting up close to the wall to mime micro-measurements, now referencing an imaginary ruler instead of a handspan.

López had traveled a great deal in his life, earning one of his degrees in Mexico and another in Canada before returning to a long career in various projects around Mexico. He knew a few things about communication and put effort into making this point clear. His voice filled up the small space, drowning out the ordinary office sounds. "Engineers do not want magnitude data," López told me. "They want to know if people should run." These professional interests had consequences for how the engineers at CIRES designed Mexico's earthquake early warning system, and the new ways they made it possible to experience earthquakes. Scientists want to know the size of an earthquake; they want devices to produce the most accurate and detailed information as possible, he explained. But that was not the kind of work that López and the rest of the CIRES team thought an earthquake early warning system should be doing.

López's way of thinking about all the choices that went into designing and maintaining an earthquake early warning system suggested how disciplinary strategies might shape the approaches to risk mitigation. The way he explained it, engineers measure earthquakes and communicate about them to people at risk in a very distinctive way—reflecting values and priorities related to their profession. In CIRES's headquarters, he was not the only person who told me about what it meant to be an engineer.

What happened at CIRES was engineering, I learned, in orientation as well as activity.[3] Certainly, some people working at the NGO designed and crafted circuit boards in carefully maintained spaces. Some visited equipment in the field and brought back damaged material for their colleagues to fix. These kinds of tasks are often associated with engineering—hands-on, mathematics-heavy work. Other people at CIRES did the kind of project management and communication work that is a necessary part of engineering but often considered a little less technical in nature.[4] So far, so ordinary. But people at CIRES also developed novel approaches to earthquake assessment that in other places might be considered scientific research. I understood from López that this was engineering, too, because of the kinds of priorities and commitments that people involved demonstrated. Further, while many people employed by CIRES had some training in electrical or communications engineering, others who were necessary to the organization's engineering endeavors held degrees in mathematics, law, graphic design, computer science, and geophysics. Some were called technicians, administrators, or designers. They may or may not have been doing engineering themselves, but they were all part of an engineering project.

The distinctive aspects of engineering projects stood out when they were contrasted to other kinds of work—specifically, to scientific undertakings. López's colleague Antonio Duran explained that members of the CIRES team were not the only people who thought that the difference between their engineering work and science was meaningful. Duran told me that the difference mattered when CIRES insights and projects were presented to the broader community interested in seismology and risk mitigation technology. He thought that the CIRES team's work was significantly undervalued precisely because of its use of engineering approaches. He compared CIRES to a donkey playing the flute. "No one notices that he plays very well," he told me. "They're surprised that he does it at all."[5] Duran's reflections complicated López's points. While CIRES engineers did sophisticated analysis and developed novel technologies to encourage people to run, Duran explained that this meant that their work was not as appreciated as it might be if they were developing scientific theory.

The fact that the CIRES team saw SASMEX as an engineering project and not a scientific endeavor played out in how they talked about earthquakes, communities at risk, and the particular possibilities that people discussed in relation to the earthquake early warning system. This was not merely a matter of focusing primarily on building and operating instruments rather than performing elaborate analyses on the data that this technology produced. It was also a matter of working with people who did not share their engineering orientation and thought about risk mitigation very differently than they did. López, Duran, and others taught me to see SASMEX as an engineer's technology. They showed me how engineering worked as an important conceptual resource both within CIRES and in relation to competing perspectives in the wider world.

In this chapter, I show how the CIRES team has relied on disciplinary strategies of engineering to frame how they address not just technology, but also environmental and social conditions.[6] When they work on SASMEX, they encounter, explore, and reflect on the physical threat of earthquakes for Mexicans.[7] They use engineering ideas to take on opportunities for technoscientific risk mitigation. I follow the CIRES team's lead to show how engineering approaches happen in context. This means exploring how the CIRES team's engineering approaches to earthquake risk mitigation should be understood in relation to earthquake hazards and scientific approaches to the same issues.

FIGURE 5.1
A technician at work in CIRES's offices. *Source*: Author (2014).

I describe how form of expertise matters for technical choices and goals related to SASMEX. I use López's explanations of what it means to measure "like an engineer" and "like a scientist," and Duran's reflections about the respect that engineering excellence fails to garner as jumping-off points to consider engineering in the context of what engineering studies scholar Atsushi Akera has called an ecology of knowledge.[8] Starting here, I can consider how engineering identity has implications for the way the CIRES team consider hazardous environments and people at risk when they make technical decisions.

ENGINEERING IDENTITIES IN PLACE

It is unsurprising that engineering logics inform SASMEX when so many of CIRES's leaders hold engineering degrees.[9] However, when I visited in 2013 and 2014, only some of the seventy-four people working at CIRES had completed training as engineers, primarily in electrical engineering or communications-related fields.[10] Of those who were not engineers, many were designated technicians and were in the process of pursuing an engineering degree. Others, including several senior consultants, had the kind of advanced degrees and academic appointments in sciences that they could have drawn on to identify as scientists, if they chose to. López was one of these. His first degree was in civil engineering and then he pursued a PhD in seismology in the United States during the mid-1980s. After a career in the petroleum industry, López had returned to work with old friends and colleagues at CIRES, where he described his work definitively as engineering.

When López and his colleagues spoke about SASMEX as an engineering project, they showcased the importance of engineering perspectives on hazardous environments, people at risk, and technologies for risk mitigation efforts. As common as the term "engineering" may be, however, what they meant by it is not necessarily obvious. Engineering identities have been a topic of some interest in engineering studies scholarship. Identities are both deeply personal and social things; and while the ways people feel and perform a given identity may vary tremendously, they are also patterned by shared experiences, circumstances, and histories.[11] Considering engineering identity in national context helps surface the particular elements of professional identity at play in López and his colleagues' risk mitigation efforts. Doing so frames their work in relation to relevant historic and contemporary politics of technoscientific practice, which may be different than those that readers assume.[12] For example, while historians have shown how engineering disciplines have developed in relation to configurations of capital and militarism in the United States,[13] drivers have taken different forms in Mexico. Historians of Mexican engineering and science instead highlight the ways that colonial interests, followed by the needs of an independent state in formation, have shaped engineering and science disciplines with attention to mining and petroleum industries as well as national development programs.[14]

As I emphasize this, I echo the many people I interviewed for this research who would not let me forget the differences between Mexico and the United States, the nation in which I was born and educated. It is true that schools in Mexico City were granting degrees while wild animals were still roaming the grounds that Harvard would be built upon. Mexico City was already an important site in New Spanish intellectual geography just decades after conquest. The European-style institutions of higher education founded under colonial rule were among the first in New Spain, cropping up as early as the mid-sixteenth century. The schools trained students in Catholic values and "useful arts,"[15] both strongly determined by the priorities of colonial elites.

The technical training fostered in New Spain focused on mining to support the extension of the Spanish empire and pay for its wars. The interests of the mining industry were more integrated into programs of higher education in New Spain than they were in much of Europe. Historians note, however, that the pedagogy that engineers and scientists of New Spain received in these programs was considered cutting-edge—simply applied, usually to mining.[16] This interest in certain particular "useful arts" that informed early Mexican engineering and science education left its mark on institutions, even as the nature of technical education changed. Technical education remained a national priority, though it was consolidated and funded in new ways after Mexican independence. During Porfirio Díaz's rule (1876–1880 and 1884–1911), industrialization started in earnest, and engineers remained in high demand. Simultaneously, because of resources mobilized by the state, education had more independence from industry. Development efforts involved engineers in post-revolutionary projects to build a modernized, independent nation.[17] In the nineteenth century, education became more accessible to the Mexican public at large. An 1857 reform made primary education both free and mandatory. The first university in North America and one of the first in New Spain was given the explicit charge of supporting both research and lay knowledge at the university level, changing its name from the Royal and Pontifical University to the National Autonomous University of Mexico, or UNAM. Founders prioritized developing a school of engineering, making it among the first offerings available in Mexican public higher education. While UNAM valued research, they housed faculty who undertook it in different university organizations (and sometimes different buildings) than those who taught professionals-to-be. There was also markedly less recognition and fewer resources available for careers in disciplines identified as science rather than engineering.[18]

While state priorities have shaped engineering training and professional practice, Mexican engineers have also had a tremendous influence on Mexican policy and research programs. Engineers from elite families like Nabor Carrillo Flores[19] and Emilio Rosenblueth[20] took on prominent roles in national and international policy in the twentieth century, as well as in their own engineering projects. They facilitated development of new institutions and fields of research in Mexico, commented on public life, and represented the nation on the international stage.

Today, engineering offers a pathway to reliable professional success, with a respectable income, for Mexico's growing middle class.[21] Perhaps this explains why engineering disciplines are attracting attention from students, capturing nearly 30 percent of those entering professional degree programs in 2011, the year I began my fieldwork in Mexico in earnest.[22] Although the career is growing, there is some concern in places like Mexico's National Engineering Institute about the quality of education being provided as training options proliferate in popular schools.[23] Even as the ranks of engineers grow, the profession retains some elite standing. It is not uncommon for Mexican politicians to earn an engineering degree before pursuing elected office.[24] The story of López's professional life—his early training in Mexico that focused on engineering, his studies abroad during Mexico's economic crisis, his subsequent work in the petroleum industry, and his return to collaborate with former classmates in his role at CIRES— all happened in the context of Mexican history. He and his colleagues did not explicitly describe engineering as the product of large-scale historical forces, however. Instead, they told me about everyday work, particularly related to designing technologies in relation to predetermined goals.

When López told me that engineers "want to know if people should run" instead of collecting information about earthquake magnitude, he was talking about engineering as a practice that concerns itself with solving urgent problems rather than producing new, theoretically robust ideas to contribute to a body of scientific knowledge. He was talking, further, about a profession that emerged in Mexico with a long-term emphasis on direct utility for addressing a given problem, for developing the nation, and for the well-being of engineers themselves. López's proud emphasis on the basic importance of determining what people should do in the seconds before an earthquake had implications for more than the application of his work, though. It influenced the way he and his colleagues understood hazardous environments, with the consequences of this information built into both the design and function of CIRES's earthquake early warning system.

HOW ENGINEERS MEASURE

Throughout this book, I have used "technoscience" and "technoscientific expert" to describe a variety of practices and people. These terms remind us that engineering and scientific objects and activities, diverse as they may be, have a great deal in common, particularly because they are inextricably embedded in the broader social and material systems in which they are produced, practiced, invented, or used.[25] While analyzing "technoscience" as a single set of ideas and practices can offer powerful insights about the world we live in and how we understand it, there is also a lot to be gained from considering how technoscientific practices form in relationship to each other. In this case, that means grappling with the implications of engineering identity and practice at CIRES, where people told me in no uncertain terms that there is more than one correct technoscientific way of understanding earthquakes, and that while the way you choose to do so is related to your disciplinary identity, your training does not alone determine how you approach a problem like earthquake early warning.

Although López holds a PhD in the science of seismology, for example, he was quite clear that in his work at CIRES his orientation toward practical knowledge and a certain form of problem-solving meant that he was working on an engineering project. Others at CIRES told me similar things—that their work was engineering, developed in light of engineering priorities, regardless of the degrees held by those involved. The identification-by-behavior went both ways. They were also comfortable telling me that people with degrees in engineering, or even in positions as engineering professors, were acting like scientists rather than engineers.

CIRES made it evident that López and his colleagues were using the same kind of careful calculation—and sometimes the same kinds of tools—that I might find in a scientific lab. SASMEX relied on equipment that they carefully maintained and updated with well-thought-out expansions. The distinctions they were making between engineering and science had to do not with tools or techniques, but with orientations toward their work and toward earthquakes themselves. I had certainly encountered efforts to distinguish between engineering and science that painted the former as a straightforward outgrowth or implementation of the latter's principles.[26] Such descriptions are reductionist and do not take engineers and engineering practice itself seriously. Engineering and scientific practices may adhere to their own social and epistemic frameworks—defining what is appropriate to do and know in distinct ways—in relation to disciplinarily relevant projects. They have their own histories and ongoing relationships within systems of industrial, state, and academic power. They feature differently in public imaginaries. Engineers and scientists in Mexico's earthquake risk mitigation community, however, also share tools, ideas, and experiences, necessarily affecting each other's work practices in the process. The distinction that López was making was happening not in the context of siloed disciplines but instead in relation to work in which both engineers and scientists were deeply involved.

In an attempt to understand this relationship better, I asked Mexican technoscientists in risk mitigation spaces—that is, scientists and engineers both—about the two fields. A technician in CIRES walked me through distinctions as we sat at a bench strewn with colorful wires where he repaired circuit boards that had been brought back broken from seismic field stations across Mexico. His colleagues moved in the background, going about their own tasks, while he explained a sort of intellectual hierarchy. Engineers were not on top; scientists were. Scientists, he explained, were intellectually powerful. They developed innovations and theories to help us understand the world in new ways. "Einstein," he remarked, "was a scientist." Engineers, on the other hand, developed practical tools. Technicians, he added, are the people who do the work. As Jacqueline Fortes and Larissa Adler Lomnitz pointed out in their study on the formation of Mexican scientific identities in the 1970s, Latin American institutions have been slow to develop scientific training compared to engineering. While science is respected, it may still be easier for some to think of it as a foreign project rather than a Mexican one.[27] Many of the engineers and scientists I spoke with were more comfortable with a definition of "scientist" that could be an ordinary job undertaken by Mexicans. They did, however, emphasize the difference between making work relevant in terms of explanatory theory or direct practice. Some went so far as to compare the two approaches with jokes that illuminated, if nothing else, more of the key ideas that people used to understand engineering and science.[28] One engineer wrote an explanation out for me on a napkin over lunch that a scientist echoed in his office during an interview—engineering was science plus common sense. It became a simple addition problem: $E = S + CS$. Alternately, engineering, without common sense, was science: $E - CS = S$.

Here, "common sense" referred to trained orientations toward the practical—certain ways of understanding how work relates to, and fits within, other systems and conditions. While scientists might develop new knowledge about vast geological systems, their theories were without "common sense" if they could not be directly applied in practice (ideally at low cost). Scientists were interested in making the best possible choices in accordance with current scientific knowledge, with little interest in the practical limitations or choices that would deliver "good enough" results. Meanwhile, while engineers might work in technologies, they did not focus their energies on conceptual exploration. The "common sense" of engineering is often integrated into social and economic systems in highly conservative ways that have inspired significant critique. STS scholar Gary Downey has described the production of this common sense as a matter of defining issues into, and out of, engineering problems. Through this process, certain forces or circumstances become available for potential intervention and transformation, and others are simply taken as given conditions in the world.[29]

DEVELOPING AN EARTHQUAKE EARLY WARNING SYSTEM LIKE ENGINEERS

López explained that measuring "like engineers" informed the whole earthquake early warning system, and this approach had been integral to the technology's design since its inception. It was an engineering system from algorithm development processes to the choices to prioritize speedy processing over accuracy. López's explanations put the descriptions of measurement and analysis that I had encountered in text, meetings, and laboratory tours in context. His comments helped me understand debates between CIRES engineers and other technoscientists over how seismicity should be measured.

Earthquakes and their effects for humans and environments may be rendered in many ways. I have described some in the preceding chapters: earthquake intensity, which describes effects experienced on a given site; acceleration, which is related to how an earthquake moves a building or the ground; and earthquake magnitude, which calculates the total energy that a rupture releases. All of these measures differ significantly with respect to what they assess and communicate. The distinction that CIRES team members like López made around measurement is a bit more nuanced. They were not drawing attention to what was measured, but instead to how that measurement was treated. Whether the seismic measurements in question are produced by observation, accelerometers, or seismometers, data can be parsed and used with more or less detail. What López calls "measuring like an engineer" is a matter of making assessments about what data should be used for, and decisions about tools and analysis processes in light of those assessments.

Anthropological work on quantification helps make sense of what López and his colleagues are doing when they associate their data decisions with their disciplinary identities. When anthropologist Helen Verran writes about quantification as a matter of "materialized relations," she draws attention to exactly the kind of situation that López and other engineers outline. Strategies for measuring and making sense of the world affirm certain priorities or commitments.[30] López's distinction between engineer and scientist was exactly this: a matter of orientation toward a problem related to how knowledge about seismicity might be used. The ways that they chose to measure earthquakes, and how they assess approaches for doing so, are related to how they understand their work in relation to the environment, society, and technology—and, at this particular moment in Mexican history, to science and scientists.[31]

Engineers designed SASMEX to collect data from accelerometers in field stations. They have often chosen accelerometers with as low resolution as is practical for the sites in which they measure seismicity. This is a cost reduction strategy sometimes, but more importantly, it is a means of limiting data processing time by collecting no more data than necessary. They analyze these data with algorithms built into computers at field stations. Based on the data collected in the first seconds of an earthquake, the algorithm distinguishes between earthquakes that are likely to be insignificant to user communities and those likely to be felt. If multiple stations detect an earthquake they deem likely to be meaningful to a user community, they transmit a message by radio relay network to a central server, which triggers broadcast warnings that then go out to dedicated radios, public radio and television stations, and, in recent years, by loudspeakers and smartphone apps.

No part of this system prioritizes producing or communicating data that local geophysicists would find useful in their own careful research. CIRES engineers preferred low-resolution accelerometric sensors that do not pick up more than the seismic information in which they are interested. The algorithms that they developed prioritize speed in data analysis over detail, and the system's public communication is a matter of simple howling alerts that convey the bare minimum information to those who might hear it— enough to indicate that an earthquake was approaching but no detail about its likely source or magnitude. Throughout the system, the engineers who designed it chose to prioritize speed of transmission (giving people more time to run, in López's words) over all other concerns, including producing detailed analyses of the earthquakes that their instruments detected. This at once provides an orienting principle to build around and demonstrates CIRES engineers' approach to making knowledge about earthquakes for publics at risk. For them, information about what the experience of the earthquake will be like is not as important to communicate as the simple fact of an oncoming quake.

This engineering approach started with SAS's very first algorithm, which allowed engineers to use accelerometric data from the first few seconds of earthquakes to distinguish between those likely to be small, of moderate size, or quite large. The algorithm was developed in the late 1980s by director Juan Manuel Espinosa Aranda (not insignificantly referred to by his employees as *El Ingeniero*, "the Engineer"). The method he used to create and refine it was unlike any kind of operation scientists might have used. Not only did the algorithm not produce detailed data about the size of the quake to come, but it was made without investment in a key scientific pursuit. This first algorithm, López told me, was developed without any explicit reference to theory. As he explained, "theory" means attempts at explanation in light of generalizable laws. Theories account for why a phenomenon happens, or a relationship emerges, in the broader context of an ordered universe that scientists are working to understand better. López, with a long career behind him, understood theory to be essential to good science.[32] Scientists might have developed algorithms based on significant amounts of data and made efforts to explain why the trends that they were identifying emerged. Engineers, on the other hand, could work with much less. The algorithm designed by engineers distinguished between earthquakes based only on data from their first seconds and included no reference to universal rules.

Espinosa Aranda began to play with these data in the late 1980s. He had a degree in electrical engineering and experience working with an earthquake sensory system for UNAM's Engineering Institute. He and his team worked from the evidence they already had to develop a mode of distinguishing between the small, medium, and large quakes that the coast could produce (see figure 5.2).[33] They did not possess a great deal of data on which to draw, maybe as few as one hundred readings from the same place, many of which they had to discard. Ultimately, they did the job with just fourteen earthquake readings.

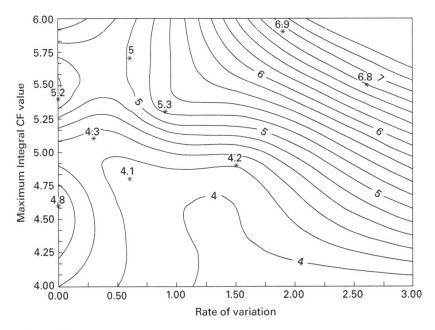

FIGURE 5.2

Magnitude calibration curves, in which magnitude is plotted against rate of variation and maximum integral characteristic function (CF). The lines designated 5 or larger are trigger criteria for SASMEX alerts. *Source*: Espinosa Aranda et al. (2000).[34]

Antonio Duran talked me through this process, explaining choices. This data set was much smaller than one that scientists might have used, but, as Duran explained, it is a very typical engineering strategy to make careful choices about data and sample some rather than analyzing it all. Using data from those fourteen quakes, engineers were able to find patterns and establish the kinds of relational rules that they could build into a circuit board: If the energy of the P wave and S wave relate as *so*, trigger an alert. If the energy of the P wave and S wave do *this* instead, do not.[35] The algorithm worked, though it took over ten seconds to produce a decision about what size the earthquake it was analyzing would probably be.[36]

As the people working at CIRES today explain, it was a sensible thing to build the system like engineers: to select sensors and develop algorithms around the needs of their project, not in light of broader knowledge. There are, however, other ways to approach earthquake early warning that respond to earthquakes not as phenomena to be better understood, but as conditions with which to scale their encounters and navigate carefully and pragmatically.[37] In other words, these engineers were orienting to earthquakes—and consequently the work of earthquake early warning—in entirely different ways than their scientific counterparts. The kinds of early warnings that engineers and scientists might produce were distinct, signaling divergent ways of engaging with threatening environments, data, technical tools, and design constraints and opportunities.

SCIENTIFIC KNOWLEDGE, ENGINEERING KNOWLEDGE

When López and his colleagues contrasted their work with "science," they were invoking an alternative vision of earthquake early warning. The CIRES team had ideas about what forms SASMEX could have taken. They could have chosen other accelerometers and arrayed them in alternate patterns across Mexico. They could have developed another kind of algorithmic analysis. This would have made, or would make, earthquakes understandable in other ways in light of differing social identities and commitments.

All the choices that CIRES engineers have made are up for debate, critique, and refinement. This is as true within the CIRES laboratories, as engineers investigate new strategies and tools that could be integrated into system components, as it is beyond CIRES's doors.[38] CIRES engineers, after all, were not the only people concerned with SASMEX and how it worked.

Emergency managers, policy makers, entrepreneurs, and scientists often had informed opinions about what might constitute an effective earthquake early warning system, though they were guided by their own disciplinary priorities. These stakeholders might articulate these interests in the context of policy spaces or in shared research communities.

We often understand engineers in relation to their ongoing, everyday practices with technologies. These practices are sometimes more or less emergent and improvisational, but they often have to do with design, use, or repair.[39] Engineering is thus often a matter of intimate encounters with technological objects. This is certainly evident in how people at CIRES explained their status as engineers. It is important to consider, however, how exactly, in López's words, "measuring like engineers" becomes a matter of knowing "if people should run," and that both technical problems at hand and the articulation of utility could be understood in alternate ways by people with disparate ideas about what knowledge is useful. Indeed, in their work on scientific identity in Mexico, Fortes and Lomnitz examine the role that ideologies about knowledge and its utility play for scientists in formation.[40] Indeed, in one controversy about how best to organize an earthquake early warning system, the kinds of knowledge a system produces and what information does for people at risk were crucial elements about which engineers and scientists disagreed.

While the debate I refer to manifested in many sites, one article provides an excellent overview. In 2007, five researchers published an article pointedly titled "The Seismic Alert System for Mexico City: An Evaluation of Its Performance and a Strategy for Its Improvement."[41] In it, they explored another way that SASMEX could work. With algorithms designed around alternate priorities and stations repositioned, their suggestions incorporated alternate principles than the engineers' designs. The article was published in the *Bulletin of the Seismological Society of America*, a well-respected English-language publication that circulates both scientists' and engineers' writing. The authors' analysis of the strengths and weaknesses of the current earthquake early warning system revealed scientific priorities, CIRES team members told me, despite the fact that several held PhDs in engineering and were employed at UNAM's prestigious Institute of Engineering.

The authors advanced two critiques of the earthquake early warning system as it stood: first, the system covered only part of the region where damaging earthquakes could originate; and second, the data analysis strategies did not produce strictly accurate results. They proposed more sensors—a suggestion that CIRES engineers could get behind—as well as a wholly distinct calculative strategy that they were more hesitant about. The strategy would produce more accurate assessments of earthquakes but would take longer to do so. It would leave a smaller window in which people could take action between a warning and an earthquake than the one CIRES engineers had built.[42] The argument in support of this latter suggestion reveals a great deal about measuring like an engineer and how scientists and other critics might perceive these efforts. In the paper, the authors compared records of SASMEX's alerts against magnitude assessments generated by geophysicists. The researchers used data related to the fifty-seven earthquakes for which the system had issued alerts over its thirteen years of operation. They found that forty-two of those quakes were not, in the end, measured at the magnitude that the algorithm had indicated. In fact, the researchers found that SASMEX's algorithm commonly issued alerts—or as they put it, "false alarms"—for earthquakes that fell beneath the rough cutoff that the engineers had designated for "moderate" quakes (greater than or equal to magnitude 5 and less than 6, and subject to a "restricted" warning dissemination in schools and emergency services agencies) and "large" quakes (greater than or equal to magnitude 6, and disseminated widely). It was, the scientists wrote, a matter of "poor performance."[43] The "failure and false alert rate is high."[44]

CIRES engineers, on the other hand, analyzed these data differently, reporting only one missed event and one false alert during the time under consideration.[45] From the engineers' perspective, the system had functioned largely as expected. Indeed, when I brought this paper up with López and others, this is what they told me. Despite some coauthors' positions in the prestigious Engineering Institute at the UNAM, the engineers at CIRES dismissed them as scientists, regardless of their disciplinary training. The system that the CIRES team had designed could operate quickly, producing information that might better be rendered as "likely to be something like a magnitude 5 earthquake" rather than "a magnitude 5 earthquake." People had time to run, and from the engineers' perspective, the precise magnitude of the earthquakes involved was beside the point. The scientists did not understand these priorities. López never said that they lacked common sense, but in his broad dismissal of the critique, it was heavily implied.

Table 5.1

The early warning system's accuracy between August 1991 and July 2004. True magnitude refers to Harvard CMT or report from Mexican Seismological Service.

Type of Alert Issued	Magnitude Estimated by the SAS that Formed the Basis of the Alert	No. of Alerts Issued	"True" Magnitude Distribution of the Events		
			$4 \leq M < 5$	$5 \leq M < 6$	$M \geq 6$
Restricted	$5 \leq M < 6$	46	27	12	7
Public	$M \geq 6$	11	4	4	3

Source: Iglesias et al. (2007).

Both explanations of the system's function are, in their own way, entirely correct; either model for system design might well work. Where the scientists and engineers diverge is in the level of accuracy that they think is appropriate for producing warnings and the size of the system that they consider necessary for the project of early warning. The CIRES team is not unhappy with their system. However, when the seismic motion for a particular event was eventually established as magnitude 4.8 and SASMEX issued an alert indicating that it was "likely to be something like a magnitude 5 earthquake," it opened space for debate about the system's efficacy.

It was not the only debate that CIRES engineers had with scientists. López and other engineers described presenting findings at geophysical conferences and having their work critiqued on grounds that they found inappropriate. The CIRES engineers were interlopers in scientific space, without theory, without accuracy. Their way of understanding earthquakes, while productive and true to their training and commitments, was simply not what the scientists they encountered seemed to require. CIRES engineers wanted to share their findings. They wanted to proceed in a way that was distinct from those that many members of the national and international community working on earthquake early warning favored. Doing so had consequences, though.

These debates over the right way to produce knowledge about earthquakes could have effects on their system's reputation. When scientists are addressing algorithmic functions and questions of alerting communication, they sometimes use value-laden terms like "failure" and "poor performance." In the institutionally precarious world of earthquake risk mitigation, some engineers voiced concerns that these efforts to assess and govern SASMEX could pose a serious threat to both CIRES's existence and that of public earthquake early warning in Mexico. In policy and scientific circles, reviews can impact the kinds of support that a technology receives. While critiques are always part of technological development, they are particularly fraught in these circumstances. CIRES engineers were concerned not only with pragmatic uses of seismic data but with pragmatic uses of critique—and they suspected their scientific counterparts had very different ideas about both.

When the article described previously was published, SASMEX was the only earthquake early warning system in Mexico, and it operated with tenuous public support. At the time, there were no backups if state sponsors decided to defund the system. Without the broad mandate for earthquake risk mitigation present in the late 1980s, building a new public system certainly was not an option. There was no easy way to supplant the leadership of an independent NGO like CIRES, and accumulated knowledge related to running such a system could not easily be passed along. More importantly, CIRES engineers were sure that they were right.

In the wake of this article and other critiques, CIRES recruited a scientific advisory board. This strategy allowed the organization to have open discussions about divergent ways that data could be handled and that the technology itself could be interpreted. This board even included some of the paper's authors. Meetings of the scientific board became friendly opportunities for scientists and engineers to share new techniques, discuss priorities, and manage how distinct ways of understanding earthquakes could relate to each other.

Engineers and scientists enacted their identities through their practices like measurement and argumentation and their orientations toward theory and practice. The disciplinary approaches were not wholly incommensurable. They did, however, emerge in the context of divergent work practices and goals, with their own ways of encountering and making sense of seismicity. Naming these orientations helped engineers at CIRES understand their system as it could have been as well as how they would prefer it not to be. It helped them consider their work and their priorities, not just as "common sense" but as a matter of particular ways of approaching earthquakes that were distinct from those that others might develop.

A WORKING ALERT

There are many possible ways to make sense of earthquakes. There are even many strategies for producing seconds of warning before an earthquake reaches a population at risk and many ways to make those seconds valuable. Invoking an opposition between "science" and "engineering" is one of the key ways that various experts in this Mexican technoscientific community parse and choose between multiple, sometimes mutually exclusive, approaches to earthquake early warning and, simultaneously, describe how they see technology and its relationship to environments in which it is embedded and communities it might serve. The stakes are high because early warnings are supposed to protect people. Debates around early warning strategy continue in many forms. They emerge in arguments about whether alerts like the ones I described in the first and second chapters actually "worked" and also in discussions of other early warning systems.[46] As the paper described earlier showcases, though, utility of one approach over another is not always easy to determine even as the distinctions between them are important to their advocates.

A conference held on the thirty-year anniversary of the 1985 earthquake in a new modern conference center on the south side of the UNAM campus was just one more place where people discussed and debated their approaches. The meeting was designed to bring engineers, policy makers, and geophysicists together. For an entire day of the two-day event, the CIRES team sat in a small room with alerting experts including a British entrepreneur, a Japanese engineer, the scientists developing a robust system for earthquake early warning in the United States, and me. Their presentations covered recent developments in their algorithms and data work and shifted between English and Spanish.

These talks were followed by questions about data derived from seismic sensors, strategies for analyzing it, and the applications of these processes, some quite direct but not unusually so for the genre. Had this kind of data been considered? Or another? Why was such a calculation necessary? In the day's conversation, sensory data and its analysis were topics of presentations. There was no other place in the conference for discussion of the issues that concern me here: the bigger-picture goals of a such a system or its use. These were rolled up into the discussion time scheduled after presentations.

Discussions about data became, inexorably, frank discussions about priorities. Dr. López, as he was wont to do, was clear. He used English, so that the people in the room who did not speak Spanish would understand him as he explained that the Mexican system was unlike their similar, still-nascent system ShakeAlert, which had largely been developed at Berkeley and CalTech and was closely integrated into ongoing scientific research. SASMEX was a matter of engineering. He told the room in his usual booming voice: "It's not very accurate. We know. Everybody knows!"[47] If people were going to ask questions about the system and how it worked, they needed to start there and see its limitations as the product of certain priorities, politics, and capabilities of pursuing some opportunities and not others.

López's efforts to explain SASMEX highlighted contrasts between the extant system and visions of critics or projects that foreigners were in the process of developing—and the ways that engineering identities are mobilized in context. His explanation of the system and its goals offered excellent insight into how he and his colleagues consider their engineering identities while attempting to change how Mexicans live with earthquakes. As I discuss in the next chapter, though, CIRES technologies can frame transformative encounters with Mexican environments and society in other ways too.

6 FIELDWORK AND NEW ENCOUNTERS

A series of flat screens mounted on walls throughout the headquarters of the Centro de Instrumentación y Registro Sísmico (CIRES) displayed digital maps of Mexico, shaded in deep green and sandy brown, as shown in grayscale in figure 6.1. Every so often, bright green and yellow rings spread across the screens. These were the P- and S- waves of earthquakes, expanding smoothly out from their epicenters through Mexican territory. As the rings expanded, they spread past vibrant dots. Some of the dots represented the field stations set up to register earthquakes, while others indicated the Mexican cities plugged into the earthquake early warning system, but the map did not focus on them. This visualization was designed solely to show Mexican earthquakes as extrapolated from what field stations register.

Every floor of CIRES headquarters contained at least one screen showing this information. The display was also available at CENAPRED, the National Center for Disaster Prevention, to the south of the city, as well as at all the other centers that processed field station signals and turned them into warnings. A small screen in a server room in Oaxaca City displayed this information, as did similar screens in other Civil Protection offices of CIRES user communities in Puebla, Morelia, Guadalajara, Acapulco, and Chilpancingo. At all of these places, screens helped people understand SASMEX as a networked system, and Mexico as a seismic nation.

The constant visual presence of the map was accompanied by other reminders of the network's activity. Every six hours, CIRES's central computer attempted to contact each field station to confirm that it was ready to register, analyze, and alert users to any potential earthquake activity.

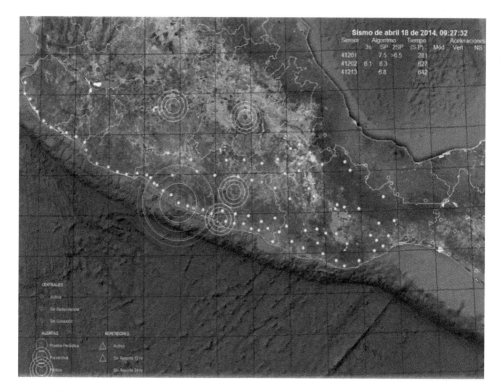

FIGURE 6.1
A map of the system showing the Good Friday earthquake of April 18, 2014. In the original, differently colored circles indicate functioning field stations, nonfunctioning field stations, and user communities. *Source*: CIRES (2014).

CIRES engineers used the ping to gauge field station function and calibrate the connections between field and center. The sound of the word "*¡reportándose!*" emanating from speakers in a masculine and mechanical voice became a regular feature of my time at CIRES offices. "Reporting!" The call would be answered by beeps and a higher-pitched, female-sounding voice listing one of the ninety-eight field station locations around the country.[1] This exchange echoed across everything that happened at the center's two main buildings, informing work and providing a backdrop to conversations.

It takes more than just managing the kind of station data visually and audibly represented in CIRES offices to keep SASMEX running. Operating the earthquake early warning system requires engineers and technicians at

CIRES to engage with aspects of Mexican environments and social worlds that may not be evident to the uninitiated. In this chapter, I argue that technicians' encounters with a variety of conditions (only some of which were visible on screens in CIRES offices) inform the success of their project. The work of going to the field to repair and update stations makes it possible—and, in many cases, necessary—for CIRES engineers and technicians to develop new ways of understanding and relating to Mexican conditions. Considering the fieldwork involved allows us to see risk mitigation work not only as an attempt to intervene on relationships between hazardous environments and society, but as a kind of activity that may change how technicians themselves encounter environments and society.[2]

In interviews, formal presentations, and casual discussion in CIRES offices, engineers and technicians told stories that helped me understand this dynamic. They discussed the environmental and social conditions that have coincided with their ability to extend the seismic monitoring system as well as the political contexts in which SASMEX has taken shape. They described long trips and close quarters; insects, animals, and plant life; bad food and flooded roads; and poverty and the pervasive effects of organized crime. These are the kinds of features that media scholar Nicole Starosielski has called the "ecologies" of communication infrastructures.[3]

As they talked about the technical requirements of the system that they use and maintain, CIRES team members also reflected on the character of Mexico itself, particularly the territory that their work requires them to get to know. Their stories about seismic monitoring demonstrate the consequences of and ongoing practices that produce Mexican territorial politics, particularly the longstanding inequalities between remote rural places and the wealth and connectedness of Mexico City. Listening to their stories, I began to understand that what intrigued me about their routine fieldwork also made CIRES engineers and technicians reluctant to take responsibility for my well-being outside of Mexico City.

In this chapter, I describe the network of seismic stations on which earthquake early warning depends and consider what the distribution of the sensors and the processes by which they are maintained reveal about Mexico.[4] The network as a whole constitutes an infrastructure (or part of a semi-integrated infrastructure as described in chapter 4). Researchers often point out that infrastructures only begin to concern most of us when they are damaged. For designers and maintenance people at CIRES, however, infrastructure requires constant attention.[5] In previous chapters, I have made the case that environmental risk management systems need to be understood in relation to the social and environmental conditions that motivate their design and use. Here, I show that operating an earthquake early warning system entails producing new knowledge about more than seismicity.[6] This work brings technicians and engineers face to face with diverse environmental and sociopolitical conditions outside of Mexico City that some refer to as "the reality of our nation."

In what follows, I use data from participant observation as well as evidence from formal and informal reports, stories, and interviews with SASMEX's designers and the people who monitor and repair the system's technical components. I describe the processes of physically siting and maintaining parts of the earthquake early warning system in places subject to storms, poverty, and rampant violence. I show how diverse forms of knowledge are required for and produced in earthquake early warning. Here, I describe how engineering knowledge production involves more than those topics that might seem straightforwardly related to technology. I suggest, in light of the meaningful experiences that technicians recount, that the purpose of a technology should not structure our expectations for and understanding of its effects.

PLANNING OPERATIONS

At CIRES's Mexico City headquarters, more than a dozen engineers, technicians, administrators, and advisors meet almost every week in fast-paced "Directors' Meetings" to address the organization's projects. CIRES leaders would pour into the conference room around noon, the most senior among them sitting around a central table, the more junior back along the walls where I joined them to take notes.

The assembled representatives of CIRES's various departments would move briskly through their updates. The senior director tasked with running the meeting had been involved in the organization from the very beginning. He kept the conversation moving. On a single day, he might ask for an update on a coworker's health before inviting a member of the nonprofit's advisory board to discuss an interview with the press, request that a department head explain his newest tweaks to an algorithm in detail, and inquire about his team's ongoing effort to stimulate regulatory reform.

He solicited information on the progress of efforts to take earthquake early warning live in a new city. He checked in about the conditions around a station in Michoacán, where violence had prevented CIRES staff from checking in on a sensor for months. Each item received only a few minutes, with operational details generally provided by more junior members of the organization and more senior staff discussing how best to incorporate this information into short- and long-term plans. Any issue related to the system, and how to keep it up and running 24 hours a day, might be discussed.

This conference room was a powerful place for CIRES staff to conceptualize Mexican territory. Anthropologist Lucy Suchman has described places like CIRES headquarters as "centers of coordination" where technoscientists enact what she calls an "orientation to problems of space and time."[7] While weekly CIRES directors' meetings often included strategic conversations about data analysis and warning communication, they also entailed discussions about the minutiae that made environmental monitoring possible. Their focus on the conditions around individual stations offered a dramatic contrast to the sweeping map of Mexican earthquakes present in the room, and the simplicity of the sirens that the system produced.[8]

My trips to and from directors' meetings frequently gave me an opportunity to observe more junior engineers and technicians loading up their trucks for journeys out into the field (shown in figure 6.2). There, in a cramped parking lot within the cheerful yellow outer walls of CIRES headquarters, teams of two or three checked and double-checked their equipment. The team leaders were younger men who had been with the organization long enough to take on roles requiring independent judgment, but not long enough to achieve wholly Mexico City–based positions.[9] While they would generally try to get going early in the morning, some might still be preparing as midday approached, with one person or another running to a third-floor office space, or into storage in a spare office on the ground floor, and coming back weighed down with supplies. Since teams went to the field with detailed information about the particular condition of field stations, provided by computers during the automatic check-ins, they knew more or less which equipment they would need for each station. They performed regular maintenance but would never try to fix the more complex elements in the field—better to simply replace components wholesale and minimize system downtime.

FIGURE 6.2
Trucks at CIRES headquarters, Mexico City. *Source*: Author (2014).

The teams also knew, however, that they were likely to encounter surprises. The field was a place of unpredictable challenges, including but never limited to the earthquakes that the field stations were designed to register and analyze. Operating with an expectation of surprise, the field team members always brought along more than they anticipated needing based on the reports delivered at CIRES headquarters. The beds of their white trucks usually contained large spools of cable alongside big toolboxes packed full to the brim. The trucks were also loaded with the nontechnical things necessary to gain access to the far-flung remote stations: money for food, beds, and sometimes gifts or rent for local landowners. Teams brought along keys to open the chain-link fences that ring field stations to protect components from wildlife and thieves, big umbrellas to shade their work in sun and dripping rain, climbing gear for ascending radio towers to adjust cables or mounts on solar panels, and cameras to take photos of everything for later analysis and to illustrate their stories (as in figure 6.3). Their paths to the field were necessarily framed by territorial politics.

On their return, they would share stories and photos along with more formal reports. This display was not just for curious anthropologists. It was

FIGURE 6.3
CIRES technician working on a field station. *Source*: CIRES (2014).

also for coworkers. Whenever I would sit down with one man to talk about his trip to the field, inevitably another would join, and another. Passing teammates and coworkers would throw in a remark or two, a you-should-ask-him-about this or a jibe.

The photos that field teams showed to me and their colleagues illustrated some of the particular challenges and pleasures of their fieldwork. These images were mostly of men waving at the camera or focusing intently on colorful wires and circuit boards. Every step of their work required documentation, both for future maintenance and for simple reference points in stories. By recounting their experiences siting and maintaining stations, they let those of us who stayed in Mexico City engage with both the differences of the field from the places we occupied and the particular environmental and social conditions of the varied territories to which they traveled.

For CIRES engineers and technicians, working to manage the new ways of living with earthquakes that SASMEX makes possible has meant encountering rural Mexico and engaging in relationships with environmental and social conditions that affect the system's ability to remain operational.[10] The health of the system is inseparable from environmental and social conditions beyond seismicity. These precarious situations are often framed in terms of threats to station function and good maintenance, but stories about the field demonstrated how they welled up, crept in, or crashed down on field teams.

OUT IN THE FIELD

"It's really different when you go into the field," technician Rufio Gómez explained.[11] He has been visiting CIRES's network of field stations for a decade. The work had brought him out of Mexico City, to the country's west and south. He explained his experiences in the field as a matter of contrast. The field might contain any number of curious conditions, idiosyncratic within the confines of a given hilltop. It was, however, first and foremost a different kind of place than the busy room ringed with desks where Gómez and I sat and talked. For that matter, the field was different from Mexico City itself, the wealthiest and most powerful city in the nation. "When you go out," Gómez emphasized, "it's another thing."[12]

CIRES engineers and technicians described the territory they traveled through to make seismic monitoring possible using a language of broad distinctions. When referring to their organization's yellow-walled headquarters, a pair of converted residential buildings on a street seasonally roofed in purple jacaranda blooms in a middle-class neighborhood in Mexico City, many simply said "CIRES." A survey I conducted of CIRES employees revealed that many members of field teams, and indeed many CIRES employees, were born in Mexico City; it is home.[13] In contrast, they referred to the accelerometric stations they maintained and repaired as "the field." That distinction framed their descriptions of encounters with the changeable, diverse places they visited.

"The field" is a useful and troublesome construction familiar to many sciences. The field, for CIRES engineers and technicians, is a general category but one used in relation to an assemblage of often-remote sites in the high plains of western and central-southern Mexico, made relevant by the presence of seismic stations. For CIRES engineers and technicians, the field is the kind of place that one has to drive for hours and sleep in strange beds for days at a time to reach. It is also an all-purpose name for the territory that CIRES employees drive through to get to field stations. SASMEX depends on stations spread throughout six states, situated in different bioregions, subject to different air currents, and arranged to detect seismic energies released at different depths; but for CIRES technicians and engineers, they still comprise a single kind of place.

This elision of difference between diverse field sites reveals a political geography that places Mexico City at the center of the nation, with everything else in "the provinces." This term bears connotations of periphery and relative unimportance. It is just one instantiation of a trend I saw during my years of visits to Mexico City. Many of the professionals I spoke with described the city and the rest of the nation in this way, as very different places where very different things happen (such as encounters with big cats like those figure 6.4 warns of).[14]

Even though technicians sometimes broadly refer to the field as generally "other" to their Mexico City home, their fieldwork involves responding to a variety of deeply specific environmental and social conditions. Over the course of the early warning system's three decades of existence, its network of field stations has expanded, with sites spreading along the active

FIGURE 6.4
A sign that field teams were unlikely to see in urban Mexico City. *Source*: CIRES (2014).

faults likely to produce levels and types of seismicity dangerous to its hand-ful of user communities.[15] The conditions around these stations are change-able, presenting the possibility that field teams may encounter a diverse set of immediate challenges that defy easy comparisons to Mexico City.

Along the coast of Guerrero and Michoacán, high humidity and salt from the ocean fill the air. The mountains form a sort of barricade between tropical climates and places that can claim, at least, dry winters. Coastal storms blow in from the west in the summer and fall, and violent thun-derstorms dump water and electrical charge on field stations. The hilltop siting and metal construction of the towers, designed to facilitate line-of-sight communication between the stations, attracts lightning, which poses serious threats to the network infrastructure. The field stations, like the one pictured in figure 6.5, run on batteries charged by small solar panels, and power surges can produce false alerts or disable a station.[16]

The more southerly stations escape the coastal storms, but their moun-tainous terrain creates other challenges. Oaxaca state is also tropical, but the mountains rise to drier, higher ground. The stations, meanwhile, are

FIGURE 6.5
A field station. *Source*: CIRES (2018).

subject to some of the same slow threats that trouble those on the Pacific coast: insects and animals chew at wire casings, hives of bees or wasps take up residence on solar panels, and moisture fouls electronics. Theft, too, has been an ongoing problem for remote field stations everywhere; the solar panels are particularly attractive.

The design of field stations has changed over the years in response to technicians' encounters with these hazards. Equipment has become increasingly airtight, the solar panels smaller and subtler. Now stations run on as little power as possible, extending components' lives and reducing the need for the showy photovoltaic cells that are attractive to thieves. Some corrosion cannot be eliminated—it happens as the steel fastenings inevitably become corroded by sea salt or factory fumes in the breeze—but the functioning computers that make the systems work are protected within casings filled with gas, sensitive only to energy from their batteries, check-in requests from the system's brain (as they call the central computer in Mexico City), and the earth motion on which they run algorithmic processes. Lightning still strikes, but the stations have been reworked so that their components cannot be destroyed by electrical charge or, for that matter, the desperate monetary needs of locals who might have use for a photovoltaic cell.

Alejandro Pérez was a younger technician with only a few years on the job. Gómez drew him into our conversation. Pérez told me that sometimes, when he was up a tower working in horrible wind and rain, he felt out of place. "I could be in my house, watching TV . . . What am I doing here?"[17] But then, he said, he would realize that he left home to advance seismic safety. That might have been his motivation, but fulfilling it entailed engaging with much more than seismicity. Instead, Pérez and field workers like him found themselves interacting with the complex implications of Mexico's political geography.

FIELDWORK

"The first thing that we're confronted with in the field is access to the places where we have equipment. Because there are places that are very hard to get to,"[18] Gómez explained in one of our early conversations about his field experience. Over the years, he had plenty of opportunity to observe. Many field stories I heard were punctuated with recitations of trouble reaching stations. Roads had been shut for protests, maybe they were washed out, or perhaps they were simply in bad condition, requiring slow and careful navigation, as in figure 6.6. Before CIRES provided its employees with trucks that could handle mountain roads, workers had to hike. The remoteness of the stations informed the stories that technicians and engineers told about them, shaped the kind of work they did on site, and even determined when and if teams visited to do maintenance.

Many of the roads that led from highways to rural communities, and then eventually up to the stations, were barely maintained if at all. Access

FIGURE 6.6
A truck navigates a washed-out road. *Source*: CIRES (undated).

is difficult everywhere. To illustrate their anecdotes about speed and mobility, engineers and technicians showed me photos of unpaved roads,[19] some consisting not of rocks, loose gravel, or sand, but instead of mud, water, and boulders. While there is not a great deal that CIRES's massive white Chevrolet trucks cannot handle, almost every presentation by the field teams involves a shot of them stopped by landslides or struggling up something that might have supported a whitewater rafting trip more easily than something with four wheels.

Stations receive three to four visits for maintenance and part replacements from technicians each year. This includes visits in the wake of significant earthquakes, both to assess damages and collect data directly from field stations. The teams of young men from Mexico City move out to the field and back before and after rainy seasons to prepare stations for violent weather and repair them after it. They strive to fix any problems before they get bad enough to impact system up-time. They cultivate a practice of checking and rechecking equipment.

Field stations are not sited for convenience. Gómez's brief description of the process of setting up a station—"Measure, get permission, try to get the connection to work"[20]—did not refer at all to ease of access. Instead, his summary was all about making sure the location was appropriate in technical and legal ways. Measurements might involve earth scientists and communications specialists identifying someplace geologically appropriate, with good sight lines for radio connection and within reasonable distance of other stations. Access to the land had to be obtained legally, in line with local land laws. While CIRES added no stations while I was visiting, many people I spoke with—both those in field teams and those who now direct their activity from CIRES headquarters—remembered traveling to do site assessments in the early 2000s. After identifying and measuring the sites, the second process started, Gómez said: "One has to solicit permission."[21]

When a proposed field site is near an existing community, field workers take that into account. CIRES teams used to bring people from Civil Protection—officials employed by state agencies—with them on their trips to potential field sites. They soon realized that they would have better luck on their own. Civil Protection agents change from government to government, and communities who have been promised infrastructure or food subsidies tend to find that those promises go unfulfilled when the administration changes. They have a bad reputation, and the CIRES teams did not want to be associated with those governmental failings.

Working on their own, field teams would establish contact with local leaders, identify landowners, and make deals. That takes time and effort, too. People with claims to the land may not trust wealthy visitors from Mexico City with big plans. Gómez recounted the experience of siting stations in communities in the Oaxaca mountains who were concerned about exploitation. He focused their attention on a nearby community that used the earthquake early warning system rather than distant Mexico City. "Everything that we're doing is for the City of Oaxaca," he would say, "so it's not for your town, here, but perhaps one day you'll visit Oaxaca or have a kid that studies there, or another family member there. And for those people, who are in Oaxaca, this will be of service to them."[22] He understood their hesitation, he told me, and tried to help them see the system as something that could somehow benefit them or their loved ones.

CIRES's agents might offer to pay a small fee to soften a landowner's or community's disposition.[23] The organization would often also fund a project to improve its relationship with locals: perhaps computers or projectors for classrooms, or furniture for school administrators. In some places, they would contract a local official to keep an eye on the station. The engineers and technicians I spoke with were straightforward about the extra pesos they had given to facilitate friendly relationships, framing them as contributions and acts of good faith, though they understood that the money could seem like bribes.

Once a station is sited, field teams work with local contractors to pour concrete and erect the tower. But field workers' relationships with communities do not end with construction. If someone on-site had been left responsible for the field station, they would often hold the key to a station fence. For each maintenance visit, CIRES teams would have to locate the keyholder, identify themselves, and show ID to access the stations, a process that could take hours.

Some communities renegotiate once stations are already in place, suddenly demanding monthly rent for access to a field station. "First the money, then you go up. If there's no money, then you can't work,"[24] Gómez and Pérez said they had recently been told. At the time, they did not have the funds, but they left and then returned with the amount that community leaders had asked for. They told me that paying a few thousand pesos to access the site was a relatively small amount in the grand scheme of things. The station in question was already sited, and it would have been no small issue to neglect it. The system requires that stations stay up and running. Gómez and Pérez returned with funds from CIRES, paying backlogged fees that had accrued as they did, thereby maintaining a good working relationship to the extent possible.

Field teams are deft drivers, ambassadors, and negotiators. They solve access problems so that they can set up and maintain stations. While they do so, their efforts become part of local economies, local ecologies, and local semiotics. Some of this the field staff expects, but some of it takes them by surprise.

PART OF THE LANDSCAPE

SASMEX field stations have become such an ordinary part of the landscape that people have begun to use their recognizable red and white stripes and chain-link fences for projects that have nothing to do with earthquake early warning. Within the borders of Michoacán and Guerrero, on hilltops within range of access roads, CIRES field teams learned about radio towers stretching up to the sky that look like earthquake early warning field stations. These towers boast fences and tiny photovoltaic panels and sit on concrete bases with inset locked metal boxes filled with instruments. No matter how much they look the part, though, none of the keys that field teams carry work in their locks, nor do CIRES trucks carry replacements for their contents.

These towers were, by all reports, remarkably convincing simulacra of the seismic sensing field stations that CIRES had built across Mexico. The boxes were placed just so, the towers painted carefully. But they did not belong to SASMEX. There were small differences, though. Boxes that should open were welded shut. Instead of housing seismic sensory works, they held communication equipment; at least, this is what concerned community members told CIRES field teams. Fake field stations dot the landscape in the rural areas not entirely under the control of the Mexican federal government. The devices presumably facilitate communication between members of organized crime or paramilitary groups in the mountains where there are no cell towers to be found.

Alejandro Pérez, one of the technicians I spoke with regularly in CIRES headquarters, told me the story. He had been giving a talk for community members at the Infrenillo station in Michoacán when someone asked his team to demonstrate station maintenance routines. They did, gamely, opening the fence first and then the locked metal box of seismic sensory equipment to show off the circuit boards that CIRES builds in its Mexico City offices. When locals saw the equipment, he says, they were surprised. It was different, they told him, from what they had seen when an army sergeant destroyed a station that looked like it belonged to CIRES with dynamite. Out of the rubble of this destroyed station had come very different batteries and transmission equipment. "That's how we knew," he said, "that those guys were cloning our field stations."[25]

Pérez calls the copycat stations "clones"[26] and assumes that they are related to organized crime—a safe assumption based on the available evidence.[27] Their very presence indicates something about organized crime in rural Mexico: its ability to hide in plain sight. Organized crime is a threat that Pérez and his coworkers must consider when they go into the field. When they know that a region is violent, this knowledge keeps them from maintaining and repairing field stations entirely. In other locations, though, the potential for organized crime is simply one concern among many.

In 2014, while I visited Mexico City, the field held particularly significant dangers for the engineers and technicians who ventured out from CIRES headquarters. That September, forty-three student teachers had vanished while traveling through Guerrero state, bringing renewed national and international attention to the central role that organized crime had taken on in many Mexican communities.[28] While the "Ayotzinapa 43" were especially well covered by the news media, the Mexican government counted more than twenty thousand murders during the year that I worked in CIRES headquarters—lower than the national murder rate in 2011, but still more than double that reported in 2007.[29] Guerrero and Michoacán are among the states made most dangerous by organized crime, and together they hold nearly half of the seismic monitoring system's field stations.[30] These places were truly dangerous for field teams, but these conditions were inconsistent and changeable. While CIRES directors needed to keep as many sites as possible active and ready to detect earthquakes, sometimes they could no more send engineers and technicians to visit field stations in contested territory than they could send them out into seasonal storms.

By the time CIRES heard about it, the tower in the Michoacano community had come down. Others, Pérez was sure, were still out in the mountains, a sort of semiotic parasite living on the official image of the CIRES field stations, taking up residence in the system of rural sites as if it were just another one of them.[31] The field stations, developed to sense earthquakes, also respond to and change the social and physical environment of rural Mexico. They offer camouflage from authorities or rivals and provide shelter for animals, income for landowners, or seismic readings for CIRES. Even as they are set up to resist the corrosive air, economic struggles, and lightning strikes, field stations and CIRES have become productively integrated into the environments they occupy.

The risks in the field have the power to do more than surprise CIRES teams. Pérez's coworker, walking by, told me that he had been exploring in Jalisco and came upon what he called "a hidden narco landing strip." He and his coworker fled. There were many situations like this, both explained, in which the teams simply did not know "exactly what we were getting into and with whom we were getting into it."[32]

CIRES directors made decisions about going to fix stations or leaving them be by trial and error. "There's no manual, no guide that can tell you, here it will be like this or here it will be like that," senior technician Rufio Gómez explained. CIRES developed a station protocol to take evidence of systemic problems into account, but when it came to organized crime, "evidence" might be slippery—a feeling that a team member got from an armed soldier, or perhaps a story of exploitation from a community member rather than something they could photograph and document for reports or friends at CIRES headquarters. "This is how we indirectly realize how things are," Gómez said.[33]

BACK FROM THE FIELD

When they return from the field, CIRES engineers and technicians put their equipment away. They make reports and tell stories, while their bright rain jackets dry on their trucks' side mirrors (as illustrated in figure 6.7), some evidence of their travels evaporating away while other kinds are circulated. They incorporate their detailed and informal reports into plans to tend to some repairs, to redesign equipment, to visit stations, or to leave them out of commission until local violence has subsided. On their field trips, CIRES engineers and technicians produce knowledge about Mexico. Although their work is explicitly about seismicity, doing it well requires that technicians encounter and understand Mexico in new ways.

FIGURE 6.7
Raincoats on trucks. *Source*: Author (2014).

Like everyone else in CIRES headquarters, I kept an eye on the simple digital maps of seismic Mexico and its glowing dots and sweeping lines. I listened to stories of the field, each with their own idiosyncratic surprises. I used the stories and documentation that the engineers and technicians produced to understand what Sistema de Alerta Sísmica Mexicano's seismic monitoring project could entail—what kinds of activities it required, and what kind of knowledge it produced above and beyond that related to seismicity.

Off the coast, the Cocos and North American tectonic plates continue to move against each other. When they slip and release massive waves of energy, SASMEX will frame a very simple message to the CIRES office and its users about an oncoming earthquake of significant size.[34] As engineers and technicians maintain this system, they experience what arrangements of Mexican power mean in what they call the "field." Their practices make knowledge necessary to seismic monitoring, but their new encounters and understandings of Mexican territory might have other kinds of impacts in their lives.[35] The young men I interviewed suggested that this work changes how they understand Mexico. These effects, however, are less documented than Mexican seismicity or the effects of storms on field stations.

CONCLUSION

Throughout this book, I make the case we should think about efforts to design new risk mitigation technologies like earthquake early warning systems as attempts to transform relationships between people and the dangerous environments we live in. I develop this case as I make three arguments: that the development and use of SASMEX is more than a sociotechnical project, and has to be understood in terms of environment, society, and technology; that the system has been unfortunately marked by techno-solutionist models for risk mitigation and a lack of reliable institutional support; and that its logics are shaped by disciplinary approaches distinctive of engineering in Mexico. Considering these aspects of SASMEX allows us to understand environmental monitoring for risk mitigation holistically.

With its ethnographic approach, this book has been able to explore environmental monitoring and risk mitigation in ways that other accounts (especially those focused on multiple systems, those that evaluate isolated incidents, or those concerned with only certain technology performance) cannot. Systems like SASMEX are never simple technical tools that either succeed or fail at warning users before earthquakes. The more we approach such technologies as part of life for users and operators alike, the more we may understand what these tools do and how they do it. In this book, a close examination of SASMEX is an opportunity to understand earthquake risk mitigation as an international preoccupation and a Mexican phenomenon; as part of global trends in technoscience emerging from particular communities and environmental conditions.

That specificity is important. I have shown how this novel risk-mitigation technology developed in light of ongoing environmental conditions and disastrous events: informed by technical, conceptual, and political trends, and shaped by decisions made by specific people. This approach allows me to showcase the complex interplay of social and material worlds that include and exceed "sociotechnical" models of technology, and oppose simple environmental determinist logics for risk mitigation and disengaged, apolitical narratives of technoscientific development. We need to do better if we want to understand—and maybe even succeed at developing—environmental monitoring and risk-mitigation technologies.

While I remain guardedly optimistic about earthquake early warning's potential, fulfilling its promises is no small undertaking. Technocentric approaches to risk-mitigation technology are an impediment. When technoscientists are made wholly responsible for outcomes that require more kinds of expertise than those in which they are trained, when potential user communities are not meaningfully involved in alert decision-making (or even consistently taught about alert utility), when the infrastructures that support early warning dissemination are not well-integrated, and when funding for earthquake early warning is unreliable at best, it can be no surprise if earthquake early warning falters. If, on the other hand, experts in communication and social research are brought in from the very beginning, education and community involvement are made high priorities, efforts at warning are controlled and coordinated, and funding is reliable—well, earthquake early warning might help protect people and property in the ways that advocates want it to.

These suggestions are borne out of close study of SASMEX, but they are applicable for different projects. At the very least, any earthquake early warning system can benefit from these insights. These systems vary tremendously, too, whatever their parameters. Some systems are robust, offering wide coverage and multichannel communication optimized for a variety of functions both automated and social; this is certainly true of the growing US system.[1] Similarly, Japan's earthquake early warning system transmits messages to emergency management organizations, media, and the public in a number of ways. It is designed to have many different effects, from prompting protective actions, to controlling automated functions in factories, to slowing down elevators.[2] These are far from the only other examples of earthquake early warning systems for which a book like this might be pertinent. In South Korea, public earthquake alerts are broadcast to all cell phones simultaneously across the country.[3] In northern India, a network of eighty-four accelerometers produces an alert that sounds sirens in student dormitories and for emergency managers.[4] Turkey has a small warning system that provides alerts specifically to a gas utility and the Marmaray Tube Tunnel.[5] In Romania, warnings are used primarily by a nuclear research facility and the Basarab Bridge.[6] These systems are only the beginning. I am aware of other systems that are operational or in development in China, Taiwan, Italy, Switzerland, Hungary, Israel, New Zealand, Chile, Costa Rica, El Salvador, Nicaragua, Nepal, and Canada. There are doubtless more.[7] Despite the differences in each system's particular environmental, social, and technical conditions and configurations, they all need to put nature, humans, and technology "on the map" together, as a CIRES engineer named Antonio Duran said.[8]

The practical insights I developed owe a great deal to highly academic scholarly work. It is true, as critics say, that research that focuses exclusively on developing theory may miss opportunities to lay out the real-world implications of its findings. On the other hand, studies that center narrow, practical concerns often neglect theoretical engagements that would help make their interventions more robust and thoughtful. Writing this book for multiple purposes, in light of both kinds of work, has been no small thing.

While we may often write for academics like ourselves, we do not always do so. For example, anthropology has substantial traditions to draw on for meaningful engagement with the communities we study and for whom we write.[9] STS has its own histories of engagement. Gary Downey and collaborators have used the term "critical participation" to refer to the kinds of things we produce when we take these challenges on to "make and do" outside of STS scholarly spaces.[10] STS scholar Annie Patrick points out that this kind of research on and practice with technoscientific communities is at its best when it refuses to adopt the goals and perspectives of the topics under examination—these are, after all, often part of dominant political narratives that exclude and undermine critical ideas, marginalized groups, and important alternatives.[11] Producing usable insights with and for members of these technoscientific communities under these circumstances is challenging, requiring labor above and beyond what scholars are typically trained to do.

In my efforts at critical participation, I have been particularly inspired by some feminist science studies scholarship, which theorizes moving across the boundaries of discipline and practice. In particular, Emily Martin's thoughts on "irenic"—that is, peaceful or conciliatory—kinds of engagement help me think about my research and writing as "a motivated positionality . . . in which diverse kinds of alliances could be forged with others (even natural scientists) over a variety of (changing) common interests."[12] Martin's work and the conversations that have emerged around it[13] offer me models to grapple with the perspectives and alliances that we take on when we experiment with the many vibrant modes of engagement available to us.[14] Thinking with Martin's irenics helps me resist the pernicious and seductive purity politics that often lead people in my position to consider research and writing for one professional purpose—whether that be applied or highly academic—to be by definition superior to work done for different audiences and goals. These purity politics have marked too many discussions about what makes good scholarship and practice. We must find ways to value and evaluate research within and across fields without leaning so hard upon narrow normative imaginaries. This book is, in the end, only one contribution to diverse, excellent efforts to understand environmental monitoring and risk mitigation. I researched and wrote in light of particular opportunities (described in more detail in the appendix). I could have made different choices, and, indeed, many wonderful researchers have and will in the future.

Whatever else my contribution may be to various projects and conversations, I have sought to guide all readers to understand SASMEX as a kind of experiment unfolding over decades. I have highlighted the creativity of the people who designed, maintain, and advocate for it. Their ingenuity means that SASMEX continues to develop. SASMEX, once only broadcast to radio and television stations, scattered dedicated receivers, and privately operated smartphone apps, now triggers sirens. It has been enrolled in state-sponsored apps, too. A series of dangerous earthquakes in 2017 promoted new kinds of awareness about earthquake risk mitigation and earthquake early warning in many parts of Mexico. Now, producing notification in advance of dangerous quakes is increasingly a project for publicly and privately funded systems. History shows that memory is often short and political support may dwindle, but for now they are growing. I am invested intellectually and intimately in these new configurations of environment, society, and technology. I know I am not alone in that. Whatever form your investments take, I hope this book supports your thoughtful engagement with them.

METHODOLOGICAL APPENDIX

Researching, analyzing, and then writing this book happened in three main phases spanning more than a decade. I describe these phases below in (mostly) plain language for those interested in what long-term ethnographic research and inductive analysis is like. I primarily relied on data collection methods of participant observation, interviewing, and archival research. I collected data, analyzed them, and then did it all over again. I visited and revisited field sites, key texts, and new publications. Throughout the process, I wrote.

I present my work toward the book like this because ethnographic research and inductive analysis—that is, qualitative research that illuminates how people live and make meaning—can be hard to explain to nonpractitioners. On the one hand, these methods are sometimes obfuscated or presented as quasi-mystical practices, which is misleading. Inductive research and analysis methods are as valid as deductive laboratory work, for all that we may use them to inquire into the nature of scientific validity. The precise engagements that lead to ethnographic findings cannot be recreated, but the findings themselves may be supported or undermined by other research. On the other hand, we do a tremendous disservice to the sophistication of this work when methods like interviewing or observation are considered a matter of simply asking a few questions or "hanging out." Research and analysis methods like these, if they are to yield robust results, must be understood in the context of ongoing training, mentorship, analysis, reading, and writing. All the extra stuff—the preparation, thinking, reading, talking, and reconsideration—are what help a scholar steer research design and analysis to make a project meaningful. Good ethnography requires what some refer

Table A.1
Research timeline.

2010–2012 Preparation	2013–2014 Fieldwork	2015–2021 Analysis, Reading, Follow-up, and Writing

• Background research
• Methods training
• Theoretical reading
• Discussion at confer-
  ences and workshops
• Short fieldwork trips
  for preliminary obser-
  vation and interviews
• Refining research
  questions
• Visits to pertinent
  archives in Mexico, the
  United States, and the
  United Kingdom
• Language training
	• On-site participant
	  observation at CIRES
	  offices in Mexico City
	• Archival research at
	  Mexico City collections
	  and libraries
	• Visits to Civil Protec-
	  tion officials in Mexico
	  City, Oaxaca, and
	  Guerrero
	• Preliminary data
	  analysis
	• Theoretical reading
	• Discussion at confer-
	  ences and workshops
	• Refining research
	  questions
		• Short fieldwork trips
		• Data analysis
		• Writing dissertation,
		  articles, and book
		• Further background
		  research related to
		  pertinent themes
		• Research on US earth-
		  quake early warning
		  project
		• Theoretical reading
		• Discussion at confer-
		  ences and workshops

Source: Author (2021).

to as a "theory-data-method triangle," in which all elements align to support each other. As anthropologist Diana Forsythe argues, poorly considered engagements with ethnographic methods can easily produce shallow or wholly inaccurate findings.[1] I will remind readers: there are whole libraries of careful thought on the historical, theoretical, and political forms into which this book fits. There is so much to inspire and trouble you out there. What follows are a few practical notes. If you read this, please read more.

RESEARCH TIMELINE

PHASE 1: PREPARATION, 2010–2012

My preparation for this research began with my enrollment in the University of California Irvine's PhD program in anthropology in 2010. Although I benefited from anthropological training as a student at Reed College and the University of Chicago and as a researcher at Northwestern Memorial Hospital beforehand, where I first encountered and practiced qualitative research, it was at UC Irvine where I learned more about a few research methods that have been particularly useful for this work: participant observation, in which a researcher engages in the activities that they are investigating to better understand them; unstructured ethnographic interviewing, in which a researcher guides a conversation with research participants to learn about their perceptions and experiences; and various forms of qualitative and quantitative data analysis. I also enrolled in seminars on anthropological approaches to science and technology, environment, and other key topics related to this research. These helped me understand the state of the field and introduced their important histories, supporting my ability to make wise choices in relation to methods, research questions, analysis, and representing my findings. Thinking with other students and with guidance from mentors gave me insights into my discipline and my work that I would never have developed on my own.

I came to this project with a theoretical interest, but had no pre-existing relationships to people involved in Mexican earthquake risk mitigation communities with whom I could build a project. I had to make practical connections in order to take on this project responsibly and make sure its outcomes could be useful. During my first years working on this project, I took advantage of research funding through UC Irvine to spend several months in Mexico. While I had studied Spanish for many years, an intensive summer language course and immersion were necessary to bring my language skills up to the level where they needed to be to ask questions, listen, and read widely. In this time, I also began to develop contacts, explore archives, and consider what kinds of research would be truly feasible for someone like me to perform in this place. These months of preparatory visits allowed me to develop a network in Mexico City who helped me learn about the city and the field of seismic risk mitigation. My contacts and mentors pointed me toward essential texts, concepts, people, and issues that would become crucial for my next steps. They helped me refine research questions, learning about Mexico City and the many communities involved in seismic risk mitigation work.

I used what I learned to produce short papers for academic conferences and workshops. This was an opportunity to think through new ideas and materials with colleagues and to listen to what more established scholars in related fields and working in the area were saying. This, too, gave me the chance to refine my approaches in ways that might allow me to make my findings meaningful in conversation with that of other anthropologists and science and technology studies (STS) scholars.

PHASE 2: FIELDWORK, 2013–2014

This preliminary research and training, as well as guidance from mentors, allowed me to write persuasive grants for my research. I spent twelve full months in Mexico with funding from the American Institute of Physics, the Society for the History of Technology, the University of California Institute for Mexico and the United States, and the National Science Foundation. During this time, I was invited to join a seminar on the history of science led by Gisela Mateos González and Edna Suárez-Díaz at the National Autonomous University of Mexico (UNAM). I was also guided by Sandra Gonzáles-Santos, then working at Iberoamerican University in reading and thinking about STS. Studying with these scholars helped me shape and refine my work.

During the first three months of my long-term fieldwork, my priority was to identify, collect, and review archival materials, including historical documents, journal articles, policy, and media documents regarding SASMEX and seismic risk mitigation in Mexico, especially those published since the upwelling of support for earthquake science after the 1985 earthquakes. I spent a substantial amount of time at physical archives at earthquake research institutes and state agencies including the National Center for Prevention of Disasters (CENAPRED), and the libraries of the Center for Research and Higher Studies in Social Anthropology (CIESAS) in Mexico City, and combing over online archives maintained by print media outlets. Collecting and considering these documents allowed me to analyze the historical situation out of which SASMEX was developed and consider technoscientific, popular, and policy discourses mobilized to develop it.

After I settled in, I secured permission from Center for Seismic Instrumentation and Registry (CIRES) director Juan Manuel Espinosa Aranda to spend extended time at CIRES headquarters. I observed ordinary office work and meetings. I conducted a survey of CIRES employees to better understand their backgrounds and thoughts on SASMEX. I did extensive interviews with the people who worked there and the scientists, engineers, entrepreneurs, scholars, and officials that they collaborated with, which I recorded and transcribed. I sampled carefully, primarily using chain referrals by which one interviewee might recommend others, but also seeking out those who might offer counternarratives. I sought "saturation," a point at which people were no longer sharing truly new perspectives even as they recounted their views regarding new events.

Over ten months, I spent three days each week at CIRES. During this time, I wrote fieldnotes detailing what I saw and heard; exploring how it aligned, or failed to, with what I learned elsewhere. I reviewed this material for trends and performed preliminary thematic analysis on these notes and the archival documents I had collected after I had been in Mexico for approximately eight months and visiting CIRES for around five. It was at this point that I began to note and then trace how, for example, risk mitigation was being treated, or how engineering identities were discussed and deployed in the interactions I observed.

During the last three months of my yearlong stay in Mexico, I focused on investigating the broader social and political context of CIRES's risk mitigation work. I supplemented my visits to CIRES with visits to the offices of officials and emergency managers in Mexico City. I traveled to Guerrero and Oaxaca states to learn about SASMEX and risk mitigation and gain a sense of how Mexico City compared with other places that are subject to frequent and dangerous seismicity.

PHASE 3: ANALYSIS, READING, FOLLOW-UP, AND WRITING, 2015–2021

Analysis, reading, follow-up, and writing are deeply entangled. Anthropologists choose to write material that expresses things we find in our data that are meaningful in scholarly conversations. This phase requires searching out and evaluating studies and pertinent material after research in much the same way as when we begin crafting our research questions in the first place.

To do this work, I returned to the United States: first to Long Beach, then to central Pennsylvania, back west to San Diego, and finally to Denver, Colorado. I had support during this time from UC Irvine's Newkirk Center for Science and Society and teaching positions at UC Irvine, California State University-Long Beach, and Bucknell University. I secured a post-doctoral position in engineering education at the University of San Diego and then took a faculty role at the Colorado School of Mines. These challenged me to read and think bigger even as I used thematic analysis techniques to trace relevant concepts through my fieldnotes, interviews, and the many documents I collected and wrote. I completed a dissertation and began to publish.

Meanwhile, I continued making follow-up visits to Mexico. A visit to Mexico in 2015 gave me firsthand experience with an earthquake early warning event. In 2017, tragic earthquakes gave this research a new and worrisome context to consider, and I made another visit. I began, further, to work with the US-based earthquake early warning community and to teach engineers and applied scientists just starting in their careers. Ongoing conversations with members of the ShakeAlert community and the ShakeAlert Social Science Working Group have helped sharpen my understanding of earthquake early warning practitioners' concerns and how social scientific insights might be useful to projects in this area. This fieldwork and experiences in my professional life drove me to write this book in ways that emphasized insights for application.

RESEARCH METHODS

I have built up my research with ethnographic questions in ways informed by the places and materials I have been allowed access to. My fieldwork for this book fell into four broad types: observation, interviews, archival work, and survey research. I collected data and then visited and revisited it, analyzing in light of the themes that I came to understand as crucial. I considered my notes, transcripts, documents, and survey data together to help me understand events and their context.

OBSERVATION AND PARTICIPANT OBSERVATION
Observation and participant observation are key tools of ethnographic research, giving researchers the opportunity to systematically observe how people go about their lives. I chose this method of research because it could help me learn about life with earthquakes and earthquake early warning in ways that reports highlighting specific events or incidents never could. I was interested in learning about ordinary life as well as remarkable events that may or may not be scheduled. Some of this I could investigate about by taking part myself. Some things I could not—for example, participating meaningfully in fast-paced CIRES director's meetings or fixing equipment were not within my skillset. Nevertheless, I could be an observer.

UNSTRUCTURED AND SEMI-STRUCTURED INTERVIEWS
Unstructured or semi-structured interviews are excellent ways to elicit reflection and explore themes of mutual interest in collaboration with interviewees. The fifty-one formal interviews that I conducted for this project took place in Spanish or, occasionally, English, and were largely focused

on issues that the people kind enough to talk to me were professionally involved in. They were experts, well versed and often very passionate. Asking them about their work, and the events and relations entailed, allowed me to learn more about the parts of earthquake risk mitigation that they might leave out of narratives that they published themselves but that they nevertheless found essential. Further, I could both enlist them in inquiries into puzzles that I had identified and follow them into key issues that they wanted to tell me about. This meant that an interview that I began with questions about an earthquake event or the development of SASMEX could jump to other themes, as an interviewee made offhand mention of a concept or theme I had heard elsewhere. At a reference to "engineers' work" or a "culture of prevention," I could shift gears and ask: "Wait a minute, can we talk a little about *that* before we move on? I want to know more!"

ARCHIVES AND LIBRARIES

Spending time at archives and libraries was incredibly useful for this project. Of course, the utility of archives depends on the data that are generated and collected in the first place. I found that my investigations into earthquake risk mitigation were well served by diving into materials others had assembled. CIESAS and CENAPRED collections held reports, dissertations, public outreach materials, conference proceedings, and peer-reviewed publications. Accessing scholarly books and popular press articles also proved essential to my project. I visited each repeatedly, sometimes for periods of months, to work my way through their collections. Throughout my entire year of fieldwork, I made weekly sojourns to the Iberoamericana University library to read through books while carefully sipping coffee. My visit to the Royal Geological Society in London was far too short in comparison—only one day!—but still allowed me to track down key materials to corroborate details and better understand stories that had been told to me in interviews.

SURVEY

Surveys can be very useful complements to the other methods mentioned here. They can provide a means of situating qualitative findings in context. In this case, I administered a short, anonymous questionnaire designed with the help of collaborators at CIRES. While CIRES had kindly allowed me to spend time at their offices, they were not able to share employee information with me; employees themselves had to choose to do that. A survey was an excellent tool for investigating demographic trends at the NGO: where people were born, where they studied, how long they had worked at CIRES, and so on. I used this opportunity to ask people when they had learned about earthquake early warning and what they thought of it, too. Only thirty-two people participated—that is, less than 50 percent of CIRES employees. People in some departments were more likely to do so than others, so it was not a representative sample by any means. I could nonetheless learn a great deal from those who did respond. I used simple descriptive statistical analyses to identify trends among respondents rather than search for correlations between their answers to my questions. This approach to survey analysis trends bolstered my understanding of CIRES and the people who work there. In this book, I do not report these findings in great detail, but instead use survey data to support qualitative analysis when appropriate.

ETHICAL RESEARCH

In this research, ethics begin with the guidelines that I established with the Internal Review Board (IRB) at UC Irvine, but do not end there. Typically, IRB guidelines stress the importance of making sure the appropriate people are invited to participate in research like this, that those who do are not made unnecessarily uncomfortable or put at risk, that they have freely consented to participate after having been informed about what their participation entails, and that they feel able to opt out of participation at any time without penalty. Addressing these considerations is not the same thing as engaging ethically in research, but doing so can be a place to begin and a way to remember commitments. Ethical work is whole-project work; it is involved in research design, research practice, and communication of findings. There are topics that it is more and less appropriate for me to ask questions about and then describe here. My status as a white woman from the United States, my lack of pre-research relationships with the people whose experiences I describe here, and my disciplinary identity as an anthropologist rather than an engineer, policy maker, emergency manager, or physical scientist inform my ethical responsibilities and accountabilities.[2]

When I write about responsibilities, I mean to signal a wide variety that exist between myself as researcher, research participants, and readers—not to mention all of those whose lives are affected by the choices that research participants participants and readers may make. The technoscientific work I describe here, after all, is not just any work. It has potentially broad consequences for public safety. We have responsibilities to each other and have power to influence each other's lives. We may choose to explicitly consider them or not, but they are very real. When I write about accountabilities, I consider a smaller set of relations: primarily those I have within my field, to my colleagues and collaborators, and to those whose experiences I study. Through accountability, my responsibilities can begin to be more than choices—they can be obligations, and failing to meet them can have meaningful repercussions.

I want to talk about this last group, people who have chosen to share their work and lives with me, in more detail here. Being accountable to them does not mean that I agree with them on all points, nor they with me. It does mean that I do my best to take their opinions seriously. In this, I understand them the same way that I understand any of my colleagues and collaborators. However, they are different in important ways. I write about them. I expose what they share with me to my analysis and then the scrutiny of my readers. That is no small thing, and I find that mechanisms for accountability that exist in my academic institutions are often clumsy for navigating these relationships. For example, it is standard practice that those who do not like a researcher's behavior can stop engaging with her. They can contact her IRB to complain. Ideally, however, accountability means something more subtle and ongoing: a researcher should be sensitive to signals from the people she is working with and use those to inform her behavior and the questions she asks long before they have to cut her out or lodge complaints. She should allow her questions and behavior to change in light of what she perceives in the field and ask the people in her field site and her scholarly community for guidance. She should be sensitive to their discomfort and refusals.

The people represented in this book have stakes in these issues, and I have relied on their help. In the early stages of my research, preliminary meetings with potential research participants and others concerned with earthquake risk mitigation allowed me to ask them questions regarding the things I was curious about. From these conversations, I refined my understanding of what research questions it might be appropriate for me to pursue. They helped me see what I might investigate and where I could go (or should not go) to do it. My primary methods of participant observation, open-ended interview, and archival research all lend themselves to refining areas of inquiry during the research process, allowing me to shift tracks when necessary.

Finally, writing and communicating findings can happen with help. Every time I have prepared anything for publication in journal articles, blog posts, or indeed for this book, I have sent preliminary copies of the texts to those who take center stage in them. They have generously guided me to clarify key issues or fix details. This way, there are built-in opportunities to gather more information, to change things when I go wrong, or to move forward with information about the implications of my research choices—at least, as long as people are able to read and comment, which may not always be the case.

The work of the people who informed this research is cited throughout this book. They themselves are described and quoted, sometimes by their real names and sometimes by pseudonyms. I developed an acceptable procedure with my patient IRB officers: generally, I attribute quotes to real people when they are on record for these statements or others like them in academic documents or in the popular press, while concealing their identity in relation to reflections that they offered to me in private and have not discussed publicly. I cannot claim that I have always done a good job at ethical research, or that I have made everyone happy in any of the choices I make in the book. My attempts at ethical research entail, if nothing else, a commitment to keep on interrogating the available methods.

THE FORM OF THE BOOK

I write in light of my own research findings, movements in anthropology and STS, concerns of professionals in fields related to disaster response and recovery, and the insights of my many readers and editors. These audiences have influenced how I have framed this book in several ways. First, they have informed my analysis. They prompt me to describe the spaces in which I did much of this research, the thoughtful reflections that people shared, and the isolated events that I experienced in ways that fit within larger trends documented by other scholars. I have been particularly interested in environments, society, and technology, and how technoscientists actively consider and attempt to remake the relationships between these not-so-separate domains. Considering how people make meaning related to environments, society, and technology are strengths of my training. I am particularly well positioned to study and write about these topics. Not only do they stand out to me but my reading and observation of practitioners in risk mitigation and the particular challenges they grapple with suggest that insights from such an approach may be of use to them. If these insights do not solve problems, they at least explain certain headaches and, I hope, let the people who have them know they are not alone.

Environment, society, and technology are what I chose to write about. However, the data I collected would make many different arguments and forms of writing possible. In this book, I choose to emphasize the ideas and histories that people there helped me learn. I refer to plenty of texts. A significant part of the social life of this topic unfolds in the popular press, peer-reviewed journal articles, conferences, and reports, and I have sought to showcase this while keeping the text as readable as an academic like me can manage. This book is a balancing act, reflecting all this as well as my embodied experiences in offices and field sites. Choosing to present all of this together means offering less of what anthropologist Clifford Geertz has famously called "thick description" here than I might.[3] For all its popularity among anthropologists, rich discussion of physical observations is just one mode of communicating anthropological research. There are a number of reasons to choose not to use it, or not to use it exclusively, within a book. My reasons for representing my findings as I do in this book have to do with accounting for various forms of practice and thought that are relevant to earthquake early warning. Meanwhile, I have also cast my engagements with a great deal of rich anthropological and STS thought into the endnotes of this book. I hope that both moves make the text legible, and even inviting, to readers not trained as anthropologists or STS scholars. Whether and how it all works remains to be seen.

In sum, this book is a product of many decisions and compromises. Things could have gone differently in big ways and little ones. It has been a joy, vexation, and honor to have the opportunity to make those decisions and develop this work.

NOTES

INTRODUCTION

1. Anticipating the location, magnitude, or timing of earthquakes with any accuracy (as defined by the National Academy of Science Panel on Earthquake Prediction in 1976) is not considered scientifically possible at the time of this writing. There remains substantial work to understand the earth's geophysical systems sufficiently for such an operation. See R. J. Geller, "Earthquake Prediction: A Critical Review," *Geophysical Journal International* 131, no. 1 (1997): 425–450; Susan E. Hough, *Predicting the Unpredictable: The Tumultuous Science of Earthquake Prediction* (Princeton, NJ: Princeton University Press, 2010); Richard Stuart Olson, Bruno Podesta, and Joanne M. Nigg, *The Politics of Earthquake Prediction* (Princeton, NJ: Princeton University Press, 1989). Some probabilistic long-term and short-term forecasting is possible, however; see James D. Goltz and Evelyn Roeloffs, "Imminent Warning Communication: Earthquake Early Warning and Short-Term Forecasting in Japan and the US," in *Disaster Risk Communication and Communities: A Challenge from a Social Psychological Perspective*, ed. Katsuya Yamori (Singapore: Springer, 2020); and E. Tapia-Hernández, E. Reddy, and L. J. Oros-Aviles, "Earthquake Predictions and Scientific Forecast: Dangers and Opportunities for a Technical and Anthropological Perspective," *Earth Sciences Research Journal* 23, no. 4 (2019).

2. Jennifer Gabrys, *Program Earth: Environmental Sensing Technology and the Making of a Computational Planet* (Minneapolis: University of Minnesota Press, 2016); see also Paul Edwards, *A Vast Machine: Computer Models, Climate Data, and the Politics of Global Warming* (Cambridge, MA: MIT Press, 2010).

3. Quote from interview. Translation by author.

4. Matthew Wisnioski, *Engineers for Change: Competing Visions of Technology in 1960s America*. Cambridge, MA: MIT Press, 2012.

5. See, for example, Hugh Raffles, *In Amazonia: A Natural History* (Princeton, NJ: Princeton University Press, 2002); Cori Hayden, *When Nature Goes Public: The Making and Unmaking of Bioprospecting in Mexico* (Princeton, NJ: Princeton University Press, 2003). I would be remiss if I failed to specifically address the formation of the field of anthropology here. There is a great deal to cite, but an introduction to this history can be found in George W. Stocking, *Victorian Anthropology* (New York: Collier Macmillan, 1987).

6. "Imported magic" is a phrase used among the Brazilian technical elite to refer to computers in the 1970s before it was repurposed to title a collection of essays. See Eden Medina, Ivan da Costa Marques, and Christina Holms, eds. *Beyond Imported Magic: Essays on Science, Technology, and Society in Latin America* (Cambridge, MA: MIT Press, 2014).

7. It can be useful to understand engineering as a domain of knowledge, set of practices, profession, and ideology, to crib from Ethan Blue, Michal Levine, and Dean Nieusma, *Engineering and War: Militarism, Ethics, Institutions, Alternatives* (Morgan and Claypool, 2013).

8. See Hiroo Kanamori, "Real-Time Seismology and Earthquake Damage Mitigation," *Annual Review of Earth and Planetary Sciences* 33, no. 1 (2004): 195–214.

9. Marilyn Strathern, "No Nature, No Culture: The Hagen Case," in *Nature, Culture and Gender*, ed. Carol P. Maccormack and Marilyn Strathern (Cambridge: Cambridge University Press, 1989); Donna Haraway, *ModestWitness@SecondMillennium.FemaleManMeetsOncoMouse: Feminism and Technoscience* (New York: Routledge, 1997).

10. Sandra P. González-Santos, *A Portrait of Assisted Reproduction in Mexico: Scientific, Political, and Cultural Interactions* (Cham, Switzerland: Palgrave McMillan, 2020), 8–9. See also Hebe Vessuri, "Global Social Science Discourse: A Southern Perspective on the World," *Current Sociology* 63, no. 2 (2015): 297–313.

11. Manuel Tironi, "Atmospheres of Indagation: Disasters and the Politics of Excessiveness," *The Sociological Review* 62, no. S1, 155.

12. See Sandrine Revet, *Disasterland: An Ethnography of the International Disaster Community* (Paris: The Sciences Po Series in International Relations and Political Economy, 2020).

13. Mexican scholars' impact on international, interdisciplinary disaster studies agendas represented in LA RED, the Network of Social Studies in the Prevention of Disasters in Latin America, https://www.desenredando.org/, stands as a marker of this; see A. Lavell, A. Brenes, and P. Girot, "The Role of LA RED in Disaster Risk Management in Latin America," in *World Social Science Report 2013: Changing Global Environments* by International Social Science Council (Paris: OECD Publishing/UNESCO Publishing, 2013).

14. See Héctor Beltrán, "Code Work: Thinking with the System in Mexico," *American Anthropologist* 122, no. 3 (2020): 487–500.

15. Leandro Rodríguez Medina, "Building Periphery: A Microanalysis of Subordinating Objects as Epistemic Technologies," *Sociológica* 29, no. 83 (2014): 9–46.

16. Sara Pritchard, "Environmental History of Technology," in *The Oxford Handbook of Environmental History*, ed. Andrew C. Isenberg (Oxford: Oxford University Press 2014), 227–258; and Sara B. Pritchard, *Confluence: The Nature of Technology and the Remaking of the Rhone* (Cambridge, MA: Harvard University Press, 2011). See also Dolly Jørgensen, Finn Arne Jørgensen, and Sara B. Pritchard, eds., *New Natures: Joining Environmental History with Science and Technology Studies* (Pittsburgh, PA: University of Pittsburgh Press, 2013).

17. See, for example Ashley Carse, *Beyond the Big Ditch: Politics, Ecology, and Infrastructure at the Panama Canal* (Cambridge, MA: MIT Press, 2014); Chandra Mukerji, *Impossible Engineering: Technology and Territoriality on the Canal Du Midi* (Princeton, NJ: Princeton University Press, 2012).

18. Anna Lowenhaupt Tsing, *The Mushroom at the End of the World: On the Possibility of Life in Capitalist Ruins* (Princeton, NJ: Princeton University Press, 2015); Valerie Olson, *Into the Extreme: US Environmental Systems and Politics Beyond Earth* (Minneapolis: University of Minnesota Press, 2018).

19. Janet Vertesi, *Seeing like a Rover: How Robots, Teams, and Images Craft Knowledge of Mars* (Chicago: University of Chicago Press, 2015); Antonia Walford, "Raw Data: Making Relations Matter," *Social Analysis* 61, no. 2 (2017): 65–80.

20. Gwen Ottinger and B. R. Cohen, *Technoscience and Environmental Justice* (Cambridge, MA: MIT Press, 2011); Mara Goldman, Paul Nadasdy, and Matthew Turner, eds., *Knowing Nature: Conversations at the Intersection of Political Ecology and Science Studies* (Chicago: University of Chicago Press, 2011).

21. Matthew Vitz, *A City on a Lake: Urban Political Ecology and the Growth of Mexico City* (Durham, NC: Duke University Press, 2018).

22. Gary Lee Downey, *The Machine in Me: An Anthropologist Sits Among Computer Engineers* (New York: Routledge, 1998); Jessica M. Smith, *Extracting Accountability: Engineers and Corporate Social Responsibility* (Cambridge, MA: MIT Press, 2021).

23. Ethan Blue, Michal Levine, and Dean Nieusma, *Engineering and War: Militarism, Ethics, Institutions, Alternatives* (Williston, VT : Morgan and Claypool, 2013); Atsushi Akera, *Calculating a Natural World* (Cambridge, MA: MIT Pres, 2007); Matthew H. Wisnioski, *Engineers for Change: Competing Visions of Technology in 1960s America* (Cambridge, MA: MIT Press, 2012); Matthew Wisnioski, Eric S. Hintz, and Marie Stettler Kleine, eds. *Does America Need More Innovators?* (Cambridge, MA: MIT Press, 2019); Amy E. Slaton, *Reinforced Concrete and the Modernization of American Building, 1900–1930* (Baltimore, MD: Johns Hopkins University Press, 2001); Amy E. Slaton, *Race, Rigor, and Selectivity in US Engineering: The History of an Occupational Color Line* (Cambridge, MA: Harvard University Press, 2010); Cyrus

C. M. Mody, *The Squares: US Physical and Engineering Scientists in the Long 1970s* (Cambridge, MA: MIT Press, 2022).

24. Åsa Boholm, "The Cultural Nature of Risk: Can There Be an Anthropology of Uncertainty?" *Ethnos* 68, no. 2 (2003): 175 puts this well, noting that "risk" is not really treated as a phenomenon but instead works as "a cognitive frame that produces contexts that link an object of risk (a source of potential harm), an object at risk (a potential target of harm) and an evaluation (implicit or explicit) of human consequences."

25. Lorraine Daston, *Classical Probability in the Enlightenment* (Princeton, NJ: Princeton University Press, 1988); Ian Hacking, *The Taming of Chance* (Cambridge: Cambridge University Press, 1990); Mary Douglas, "Risk as a Forensic Resource," *Daedalus* 119, no. 4 (1990): 1–16; Mary Douglas and Aaron Wildavsky, *Risk and Culture* (Berkeley: University of California Press, 1983).

26. Niklas Luhmann, *Risk: A Sociological Theory* (Berlin: Walter de Gruyter, 1993).

27. Michael Power, *Organized Uncertainty: Designing a World of Risk Management* (Oxford: Oxford University Press, 2007)

28. Adriana Petryna, *Life Exposed: Biological Citizens after Chernobyl* (Princeton, NJ: Princeton University Press, 2006); Manuel Tironi, I Rodriguez-Giralt, and M Guggenheim, eds., *Disasters and Politics: Materials, Experiments, Preparedness* (West Sussex, UK: Wiley Blackwell/The Sociological Review, 2014); Scott Gabriel Knowles, *The Disaster Experts: Mastering Risk in Modern America* (Philadelphia: University of Pennsylvania Press, 2012); Vivian Choi, *Disaster Nationalism: Tsunami and Civil War in Sri Lanka* (Durham: Duke University Press, forthcoming).

29. Emily Wanderer has demonstrated how coordinated efforts around a national "vivir major," or "live better," strategy come to frame a variety of efforts to improve ordinary life in the relation to ongoing threats to state legitimacy. See Emily Wanderer, *The Life of a Pest: An Ethnography of Biological Invasion in Mexico* (Berkeley: University of California Press, 2020).

30. Anthony Oliver-Smith and Suzanne Hoffman, eds., *The Angry Earth: Disaster in Anthropological Perspective* (New York: Routledge, 1999); *Catastrophe and Culture: The Anthropology of Disaster* (Santa Fe, NM: School of American Research Press, 2002); Michele Ruth Gamburd, *The Golden Wave: Culture and Politics after Sri Lanka's Tsunami Disaster* (Bloomington: Indiana University Press, 2013); Kathleen Tierney, *The Social Roots of Risk: Producing Disasters, Promoting Resilience* (Palo Alto, CA: Stanford University Press, 2014.); Roberto E. Barrios, *Governing Affect: Neoliberalism and Disaster Reconstruction* (Lincoln: University of Nebraska Press, 2017).

31. Elizabeth Povinelli, *Geontologies: A Requiem to Late Liberalism* (Durham, NC: Duke University Press, 2016).

32. Wendy Lesser, *The Life Below the Ground: A Study of the Subterranean in Literature and History* (Boston, MA: Faber and Faber, 1987); Rosalind H. Williams, *Notes on the*

Underground: An Essay on Technology, Society, and the Imagination (Cambridge, MA: MIT Press, 2008).

33. Abby Kinchy, Roopali Phadke, and Jessica Smith, "Engaging the Underground: An STS Field in Formation" *Engaging Science, Technology, and Society* 4 (2018): 22–42; Anthony Bebbington, "Underground Political Ecologies: The Second Annual Lecture of the Cultural and Political Ecology Specialty Group of the Association of American Geographers," *Geoforum* 43, no. 6 (2012): 1152–1162.

34. Kristina Lyons, "Decomposition as Life Politics: Soils, Selva, and Small Farmers under the Gun of the U.S.–Colombia War on Drugs," *Cultural Anthropology* 31, no. 1 (2016): 56–81; Maria Puig de la Bellacasa, "Making Time for Soil: Technoscientific Futurity and the Pace of Care," *Social Studies of Science*, 2015.

35. Andrea Ballestero, "Touching with Light, or, How Texture Recasts the Sensing of Underground Water," *Science, Technology, & Human Values* 44, no. 5 (2019): 762–785.

36. Deborah R. Coen, *The Earthquake Observers: Disaster Science from Lisbon to Richter* (Chicago: University of Chicago Press, 2013); Conevery Bolton Valencius, *The Lost History of the New Madrid Earthquakes* (Chicago: University of Chicago Press, 2013); G. K. Clancey, *Earthquake Nation: The Cultural Politics of Japanese Seismicity, 1868–1930* (Berkeley: University of California Press, 2006).

37. Virginia García Acosta, "Historical Disaster Research," in *Catastrophe and Culture: The Anthropology of Disaster*, ed. Anthony Oliver-Smith and Susannah M. Hoffman (Santa Fe, NM: School of American Research Press, 2002); Sergio Puente, "Social Vulnerability to Disaster in Mexico City," in *Crucibles of Hazard: Mega-Cities and Disasters in Transition*, ed. James K. Mitchell, (Tokyo: United Nations Press, 1999); Diane E. Davis "Reverberations: Mexico City's 1985 Earthquake and the Transformation of the Capital," in *Cities of the Global South Reader*, ed. Faranak Miraftab and Neema Kudva (New York: Routledge, 2014).

38. On navigating respectful and responsible expert communication, see for example, A. Bostrom, C. J. Atman, B. Fischhoff, and M. G. Morgan, "Evaluating Risk Communications: Completing and Correcting Mental Models of Hazardous Processes, Part II," *Risk Analysis* 14, no. 5 (1994): 789–798; and Karina Landeros-Mugica, Javier Urbina-Soria, and Irasema Alcántara-Ayala, "The Good, the Bad and the Ugly: On the Interactions among Experience, Exposure and Commitment with Reference to Landslide Risk Perception in México," *Natural Hazards* 80, no. 3 (2016): 1515–1537.

CHAPTER 1

1. Giacomo Parrinello, *Fault Lines: Earthquakes and Urbanism in Modern Italy* (New York: Berghahn Books, 2015), 6.

2. See, for example, Marina Franco and Daniel Melchor, "A Year after Mexico Earthquake, Parents Demand Justice," *New York Times*, September 19, 2018.

3. Substantial research does exist on this front, and I direct curious readers to reports such as Mario Ordaz Schroeder, Eduardo Reinoso, Miguel A. Jaimes, Leonardo Alcántara, and Citlali Pérez, "High-Resolution Early Earthquake Damage Assessment System for Mexico City Based on a Single-Station," *Geofísica Internacional* 56, no. 1 (2017): 117–113; and Edgar Tapia-Hernández and J. Salvador García-Carrera, "Damage Assessment and Seismic Behavior of Steel Buildings during the Mexico Earthquake of 19 September 2017," *Earthquake Spectra* 36, no. 1 (2020): 250–270.

4. Based on the territory claimed by pre-Colombian civilizations, a strict definition of Mesoamerica would only include southern and central Mexico along with other regions in Central America. However, Wolf's way of framing the region is significant for my work, and the region incorporates most of modern-day Mexico, so the jump from Mesoamerica to Mexico here is not unwarranted.

5. Eric Wolf, *Sons of the Shaking Earth: The People of Mexico and Guatemala—Their Land, History, and Culture* (Chicago: University of Chicago Press, 1959).

6. While earthquake effects are not directly determined by magnitude, quakes over magnitude 4 are often considered moderate. Near their epicenters, they may have an intensity on the modified Mercalli scale of IV, which means they are perceptible but may not cause damage. Quakes over magnitude 7 are usually considered large. See USGS Earthquake Catalogue, December 23, 2019, https://earthquake.usgs.gov /earthquakes/search.

7. Centro Nacional de Prevención de Desastres (CENAPRED), *Atlas Nacional de Riesgos*, accessed March 5, 2018, http://www.atlasnacionalderiesgos.gob.mx/.

8. Ashley Carse, *Beyond the Big Ditch: Politics, Ecology, and Infrastructure at the Panama Canal* (Cambridge, MA: MIT Press, 2014).

9. Sara Pritchard, *Confluence: The Nature of Technology and the Remaking of the Rhone* (Cambridge, MA: Harvard University Press, 2011).

10. I make this case in the context of critical scholarship on the non-innocent production of the very concept of nature. See, for example Donna Haraway, *Simians, Cyborgs and Women: The Reinvention of Nature (New York: Routledge, 1999)*. However, I describe the anthropogenic construction of physical conditions much more literally.

11. Cf. Dyl, who argues for greater attention to the ways that nonhuman nature and the built environment structure cities, even when those forces may be concealed in rebuilding and recovery efforts. Joanna L. Dyl, *Seismic City: An Environmental History of San Francisco's 1906 Earthquake* (Seattle: University of Washington Press, 2017).

12. The Nahuatl names of these translate to Smoking Mountain and White Woman, characters in a tragic love story.

13. Geophysicist Cinna Lomnitz's research and writing provides excellent examples (e.g., Cinna Lomnitz, "A Vanishing Island: A Tentative Reconstruction of Earthquake Hazard in Mexico City," in *International Symposium on Earthquake Disaster Prevention Vol 1*, ed. Sergio Alcocer (Mexico City: CENAPRED and JICA, 2000).

14. The name itself being a European gloss for a confederation of three Nahuatl-speaking city-states.

15. Candiani writes that the Mexica lived in a "fluid landscape." She suggests that, despite elaborate hydroengineering, the communities who lived in the Valley of Mexico at the height of Mexica power struggled with and made use of flooding in different ways. Vera Candiani, *Dreaming of Dry Land: Environmental Transformation in Colonial Mexico City* (Palo Alto, CA: Stanford University Press, 2014), 24.

16. Edward B. Tylor, *Anahuac; Or, Mexico and the Mexicans, Ancient and Modern* (London: Longman, Green, Longman and Roberts, 1861), 41. Spanish soldier Bernal Díaz described the waterways and settled lakes as "wonderous sights"; see Bernal Díaz del Castillo, *The Conquest of New Spain*, trans. with an introduction by J. M. Cohen (London: Penguin Books, [1632] 1963). This kind of wonder was not unusual. As historian María M. Portuondo demonstrates, Spanish encounters with the Americas in the sixteenth century were structured by biblical and classical narratives and by a drive to develop new frameworks to understand what they found. While her research emphasizes the epistemic and methodological work done by cosmographers, the basic insight is still useful to understand why Spaniards might see Venice or "wonderful sights" in Mexico. See Maria M. Portuondo, *Secret Science: Spanish Cosmography and the New World* (Chicago: University of Chicago Press, 2009).

17. See Candiani, *Dreaming of Dry Land*. There are certain echoes here to other colonial efforts to regulate water and rewrite landscapes, such as those described in Timothy Mitchell, *Rule of Experts: Egypt, Techno-Politics, Modernity* (Berkeley: University of California Press, 2002). On the particular logics of Spanish expertise, see Portuondo, *Secret Science*.

18. These data can be found in Shri Krishna Singh, Enrique Mena, and Raúl R. Castro, "Some Aspects of Source Characteristics of the 19 September 1985 Michoacán Earthquake and Ground Motion Amplification in and Near Mexico City from Strong Motion Data," *Bulletin of the Seismological Society of America* 78, no. 2 (1988): 451–477. Describing the severity of the lake zone's effects has inspired more than quantitative language. In 1936, for example, at the first international meeting on soil mechanics, José A. Cuervas offered Mexico City's soil up as a rare case. It was "hyper reactive," and he wrote that it offered "a field exceptionally rare to study Soil Mechanics and Foundation Engineering at large." See José A. Cuervas, "The Floating Foundation of the New Building for the National Lottery of Mexico: An Actual Size Study of the Deformations of a Flocculent-Structured Deep Soil," in *Proceedings of the International Conference on Soil Mechanics and Foundation Engineering* (Cambridge, MA, June 22–26, 1936), 301, document in the collection of Harvard University Archive). It held a great deal of promise for him and his new science, so much so that in a discipline that was pushing to systematize and make scientific the evaluation of soils, he resorted to not only descriptions of material structure and water in the subsoil but also evocative language about the "marvelous twilights" that the fine volcanic particulate would have created before it had settled to the earth.

19. Mario Ordaz, Roberto Meli, Carlos Montoya-Dulché, Lorenzo Sánchez, and Luis Eduardo Pérez-Rocha, "Data Base for Seismic Risk Assessment in Mexico City," in *Recent Research in Japan and Mexico on Earthquake Risk Mitigation* (Mexico City: CENAPRED and JICA, 1993); Hortencia Flores-Estrella, Sergio Yussim, and Cinna Lomnitz, "Seismic Response of the Mexico City Basin: A Review of Twenty Years of Research," *Natural Hazards* 40 (2007), 357–372.

20. Earthquakes signal shifts in history and violent transition in many Prehispanic Mesoamerican traditions (see Wayne Elzey, "The Nahua Myth of the Suns: History and Cosmology in Pre-Hispanic Mexican Religions," *Numen* 23, no. 2 (1976): 114–135; and Robert L. Kovach, *Early Earthquakes of the Americas* (Cambridge: Cambridge University Press, 2004). Perhaps the most famous belongs to Nahua-speaking Mixtec peoples (often called Aztecs) who, in the Late Postclassic period (1300–1519) defined movement through time in terms of creation and destruction of worlds or "Suns." While the principle of "motion" (*ollin*) created the present Fifth Sun, the motion of the Earth (*tlalollin*) would destroy it. This is described in postcolonial "Legends of the Suns" and in precontact documents. See, for example, Anne S. Dowd and Susan Milbrath, eds., *Cosmology, Calendars, and Horizon-Based Astronomy in Ancient Mesoamerica* (Boulder: University of Colorado Press, 2015).

21. Garduño-Munroy has proposed a way of understanding how earthquakes are represented in the Codex Telleriano Remensis according to a scale of seismic intensity. He argues that Mexicas assessed earthquake effects quantitatively. V. H. Garduño-Monroy, "A Proposal of a Seismic Intensity Scale Obtained from the Nahuatl Codex Telleriano Remensis," *Arqueología Iberoamericana* no. 31 (2016): 9–19. Whether or not this interpretation proves well founded, contemporary researchers find it very likely that Mexica people, among others, systematically documented earthquake qualities. Virginia Garcia Acosta, Rocío Hernández, Irene Márquez, América Molina, Juan Manuel Pérez, Teresa Rojas, and Cristina Sacristán, "Cronologia de los sismos en la Cuenca Del Valle de Mexico," in *Estudios sobre sismicidad en el Valle de Mexico*, ed. Sergio Alcocer (Mexico City: Departamento Del Distrito Federal Secretaria General de Obras, 1988); T. Rojas-Rabiela, J. M. Pérez-Zavallos, and V. García-Acosta, *Y volvió a temblar* (Mexico City: Centro de Investigaciones y Estudios Superiores de Antropología Social, Cuadernos de La Casa Chata, 1987), 201.

22. García Acosta and Suárez Reynoso show records indicating damage to the precolonial built environment related to quakes, Virginia García Acosta and Gerardo Suárez Rayunoso, *Los sismos en la historia de México vol. 1* (Mexico City: Universidad Nacional Autónoma de México Press, 1996). The quake in 1475 was referenced in Codex Mexicanus and Codex Aubin, in Torquemada's writings, and elsewhere. They note that important records are available in Joaquín Vélazquez de León, *Joaquín Vélazquez de León y sus trabajos científicos sobre el Valle de Mexico* (Mexico City: Prensa de Universidad Nacional Autónoma de México, 1977), 262.

23. García Acosta et al., *Cronologia de Los Sísmos* has shown how time became incorporated into eighteenth-century colonial seismic reports as public clocks appeared and how earthquakes were systematically compared to each other to assess the degree of a Catholic god's anger.

24. Carlos María de Bustamante, *Temblores de México y justas causas por qe se hacen rogaciones públicas* (Mexico City: J. M. Gallegos, 1837), 5; and Juan Orozco y Berra, "Efemérides seísmicas Mexicanas," in *Memorias de la Sociedad Científica Antonio Alzante vol 1* (Mexico City: Imprenta del Gobierno en el Ex-Arzobispado, 1887), 322–324 note these bibliographic sources. There may be some doubt regarding the true length of time the earthquake lasted, but the fact that its duration was worthy of note is important. These records are widely available because of a substantial collaborative project undertaken by historians and geophysicists in Mexico to collect diverse qualitative and quantitative records of seismic activity and make them, to the extent possible, transparent and even useful to contemporary technoscientists. See García Acosta and Suárez Rayunoso, *Los sismos en la historia de México*; on the work of making 450 years of Mexican seismological history available for research, see García Acosta 2004. Virginia García Acosta, "Historical Earthquakes in Mexico. Past Efforts and New Multidisciplinary Achievements," *Annals of Geophysics* 47, no. 2–3 (2004): 487–496.

25. Juan Manuel Espinosa-Aranda, Armando Cuellar, Armando Garcia, Gerardo Ibarrola, Roberto Islas, Samuel Maldonado, and F. H. Rodriguez, "Evolution of the Mexican Seismic Alert System (SASMEX)," *Seismological Research Letters* 80, no. 5 (2009): 694–706.

26. According to J. Velázquez de León, this was bigger and stronger than any the city had experienced in the century (see *Joaquín Vélazquez de León y sus trabajos científicos*, 273, cited in García Acosta and Suárez Rayunoso, *Los sismos en la historia de México*).

27. Gómez de la Cortina (1840, 16–17) among others, cited in García Acosta and Suárez Rayunoso, *Los sismos en la historia de México*.

28. José Gómez de la Cortina, *Terremotos: Carta escrita a una señora por el Coronel D . . .* (Mexico: Impresa por Ignacio Cumplido, 1840), 16–17, cited in García Acosta and Suárez Rayunoso, *Los sismos en la historia de México*.

29. Population and Housing Census, INEGI, Mexico, 1970 and 1980 cited in Sergio Puente Aguilar, "Social Vulnerability to Disaster in Mexico City," in *Social Vulnerability to Disaster in Mexico City. Crucibles of Hazard: Mega-Cities and Disasters in Transition*, ed. James K. Mitchell (Tokyo: United Nations University Press, 1999).

30. Here I have skipped many years in Mexico City's history. I recommend Matthew Vitz's book for careful attention to other ways that environmental forces, particularly water, shaped the city during this time. Matthew Vitz, *A City on a Lake: Urban Political Ecology and the Growth of Mexico City* (Durham, NC: Duke University Press, 2018).

31. S. Robinson, Y. F. H. Franco, R. M. Catrejon, and H. Bernard, "It Shook Again—The Mexico City Earthquake of 1985," *Studies in Third World Societies* 36 (1986): 87.

32. Jacobo Zabludovsky, "Bucareli." *Por Esto! Quintana Roo*, September 13, 2010.

33. Information from the Mexican Servicio Sismológico Nacional. It was significantly smaller, as magnitude is an exponential measurement. However, after the damage of the first earthquake, its effects were still nasty.

34. It was felt in the states of Jalisco, Colima, Michoacán, Guerrero, Oaxaca, Chiapas, Mexico, Puebla, Hidalgo, and Veracruz.

35. Report by Dr. Mete Sozen, professor of civil engineering at the University of Illinois to a US Senate Commerce Sub-Committee on October 3, 1985. See S. A. Lefomex, *A Report on Mexico City's Earthquakes of September 19th and 20th, 1985* (Mexico City: Lefomex, 1985).

36. State agencies and their affiliates give a range of numbers. Six thousand is one number circulated by the Mexican state. See Rubem Hofliger, Olivier Mahul, Francis Ghesquiere, and Salvador Perez, *FONDEN, el fondo de desastres naturales de México—Una reseña* (Washington, DC: World Bank Global Facility for Disaster Reduction and Recovery, 2012). Cruz Roja Mexicano has estimated fifteen thousand (reported by Mónica Achundia, "A 26 años del sismo, cifra oficial: 3 mil 692 muertes," *El universal*, September 19, 2011). For its part, CIRES offers a number of twenty thousand, following the news outlet Diario Amanecer (see CIRES, "The Earthquake of September 19, 1985," accessed December 23, 2019, cires.mx/1985_in.php.

37. Elena Poniatowska, *Nada, Nadie: Las voces del temblor* (Mexico City: Ediciones Era, 1988); and Russell R. Dynes, Enrico L. Quarantelli, and Dennis Wenger, *Individual and Organizational Response to the 1985 Earthquake in Mexico City, Mexico* (Newark, DE: Disaster Research Center, 1990).

38. Quote from interview with the author.

39. Mexican journalist Carlos Monsiváis communicates about this cogently. See Carlos Monsiváis, *"No sin Nosotros:" Los días del terremoto 1985–2005* (Mexico City: Ediciones Era, 2005). So do the voices recorded in testimonios such as Poniatowska, *Nada, Nadie*; Guadalupe Loaeza, *Terremoto: Ausentes/presentes 20 años despues* (Mexico City: Editorial Planeta, 2005); Adolfo Montiel Talonia, Juan Manuel Juarez Cortes, Luis Muñiz Fuentes, Victor Hugo Islas, Ricardo Blanco Velazquez, Jose Santos Navarro, and Evaristo Corona Chavez, *Septiembre 19/18, 7:20: Terremoto!* (Mexico City: La Prensa, nd); and Leslíe Serna, *Aqui nos quedaremos . . . !: Testimonios de la coordinadora unica de damnificados* (Mexico City: Universidad Iberoamericana Press, 1995). In their assessment, disaster researchers Dynes, Quarantelli, and Wenger suggest in *Individual and Organizational Response* that this is a product of ad hoc organization rather than the complete chaos or absence of response that some write about.

40. A detailed event timeline compiled in Dynes, Quarantelli, and Wenger, *Individual and Organizational Response* indicates that the president was quite active during these hours, but his absence from the public eye during this time was a serious gaffe. Many parts of the city did, after all, have electricity.

41. Two multidepartmental commissions were created by the president the day after the quake: the National Emergency Commission (CNE) to coordinate response outside of Mexico City, and the Metropolitan Emergency Commission (CME) to deal with the issues within it. These did not start operating until September 22, three days after the first earthquake. Dynes, Quarantelli, and Wenger report in *Individual and Organizational Response* that CME headquarters were not staffed around the clock, even then.

42. Researchers describe the complex relationship between military and civil authorities in twentieth-century Mexico. Between the 1940s and 1960s, the military and civil leadership of Mexico had been tightly integrated—with military acting in service to civilian authority. See Nora Hamilton, *The Limits of State Autonomy: Post Revolutionary Mexico* (Princeton, NJ: Princeton University Press, 1982); Judith Hellman, *Mexico in Crises* (New York: Holmes and Meier, 1983); and Roderic A. Camp, *Politics in Mexico: The Decline of Authoritarianism* (Oxford: Oxford University Press, 1999). At the time of the earthquake, there was some concern that the military might use this opportunity to accumulate power. See Adolfo Zinser, Cesareo Morales, and Rodolfo Pena, eds., *Aun tiembla: Sociedad política y cambio social: El terremoto del 19 Septiembre de 1985* (Mexico City: Grijallo, 1986).

43. Though elsewhere they did more; see Raymundo R. Palacio, "Y de repente entre escombros," in *Aun tiembla: Sociedad política y cambio social: El terremoto del 19 Septiembre de 1985*, ed. A. Zinser, C. Morales and R. Pena (Mexico City: Grijallo, 1986), 23–37.

44. It is worth noting that this is not an unusual response in times of disaster; as in A. H. Barton, *Communities in Disaster: A Sociological Analysis of Collective Stress Situations* (New York: Doubleday, 1969); Russell R. Dynes, *Organized Behavior in Disaster* (Lexington, MA: Heath Lexington Books, 1970); and Thomas Drabek, *Human System Responses to Disasters: An Inventory of Sociological Findings* (New York: Springer-Verlag, 1986). Researchers theorize that people suspend conflicts in time of natural disasters—for a time, at least. See Tierney, "From the Margins to the Mainstream?"

45. Following a period of stability and nationalization in the middle of the twentieth century—the "Mexican Miracle" national development strategy is often periodized in relation to import substitution policies and economic growth between the 1940s and 1970s—Mexico's economic situation had degraded dramatically in the 1970s. Troubled industries, subsidized by foreign loans, had been further subsidized by the IMF. The lenders assumed that the loans would be repaid with profits generated from newly discovered Mexican oil reserves. Then the price of oil crashed.

46. For an account of IMF refinancing and state response, see Diane E. Davis, "Failed Democratic Reform in Contemporary Mexico: From Social Movements to the State and Back Again," *Journal of Latin American Studies* 26, no. 2 (1994): 375–408.

47. Dynes, Quarantelli, and Wenger, *Individual and Organizational Response*.

48. Quote from an interview.

49. Sometimes to the frustration of organizers, who had to manage this wave of assistance. S. A. Lefomex, *A Report on Mexico City's Earthquakes*; and Emilio Díaz Cervantes, *Brigada plácido Domingo* (Monterrey, MX: Ediciones Castillo, 1995).

50. Díaz Cervantes, *Brigada Plácido Domingo*, 20. The Topos, or Moles, first organized in 1985. The group has subsequently become involved in international emergency response work.

51. These testimonios, true to their genre, are always political. A testimonio is a tricky genre in Latin America. As Anne McClintock defines the genre, it is a story "told to a journalist or anthropologist for political reasons," with "an implied and often explicit plural subject," making it a story with a speaker rather than a story about the speaker, incorporating into a personal narrative things that might not have happened, precisely, to the author but which are nonetheless essential to their subject position and the story they write. See Anne McClintock, "'The Very House of Difference': Race, Gender and the Politics of South African Women's Narrative in Poppie Nongena," *Social Text* 26, no. 25/26 (1990): 218. A testimonio can render experiences as at once both narratives of personal experience and representative of shared trauma. See Diane Nelson, *Reckoning: The Ends of War in Guatemala* (Durham, NC: Duke University Press, 2009).

52. Carlos Monsiváis, *No sin nosotros*, 9.

53. Corruption itself was not as vulnerable as individual politicians. Clientelist relations, so often a component of the practices indicted as corruption, are no less essential to the functioning of power, and have hardly been replaced by transparent democratic processes (see Jonathan Fox, "Governance and Rural Development in Mexico: State Intervention and Public Accountability," *Journal of Development Studies* 32, no. 1 (1995): 1–30 and Tina Hilgers, "Clientelism and Conceptual Stretching: Differentiating among Concepts and among Analytical Levels," *Theory and Society* 40, no. 5 (2011): 567–588.

54. Observers of the 1988 election find it very likely that the PRI engaged in fraudulent practices. Some believe that Cárdenas actually won, though others simply suspect that Salinas's win was by a much smaller margin than official numbers indicate. See Camp, *Politics in Mexico*.

55. Before, the PRI party had nominated leaders of the Federal District of Mexico City directly. Heather Levi recounts this transition as part of her ethnography on the political and social life of Mexican wrestling. See *The World of Lucha Libre: Secrets, Revelations, and Mexican National Identity* (Durham, NC: Duke University Press, 2008).

56. The PRI reasserted its hold in 1991 congressional elections, but in 2000 the nation's second-ranking party, the Partido Acción Nacional (PAN), gained the presidency with support from migrants living outside of Mexico's borders. See Luin Goldring, "The Mexican State and Transmigrant Organizations: Negotiating the Boundaries of Membership and Participation," *Latin American Research Review* 37, no. 3 (2002): 55–99.

57. This is the estimate published by Dynes, Quarantelli, and Wenger, *Individual and Organizational Response*, 3; though as Diane Davis points out, a great deal of reconstruction money should be understood to have been diverted to other uses, especially given Mexico's troubled economy at the time. See Diane Davis, "Reverberations: Mexico City's 1985 Earthquake and the Transformation of the Capital," In Faranak Miraftab and Neema Kudva, eds., *Cities of the Global South Reader* (New York: Routledge, 2015), 203.

58. Quote from an interview.

59. Comisión Nacional de Reconstrucción, *Bases para el establecimiento del sistema nacional de protección civil* (Mexico City, 1986).

60. It had sister organizations, Centro de Investigaciones Sísmicas (CIS) and Centro de Estudios Prospectivos (CEPRO). All were built up under the auspices of the NGO Fundación Javier Barros Sierra. The Foundation itself was named after a well-known civil engineer with a powerful influence on Mexican policy and education, and the safety-related missions of these research organizations resonated strongly with this legacy. However, of all of them, only CIRES remains; CIS, CEPRO, and even the Foundation Javier Barros Sierra were dissolved.

61. Union Geofisica Mexicana, "Declaration of Morelia," *Excelsior*, November 27, 1986.

62. Geophysicists are no longer sure that the Guerrero Gap will produce a big temblor. It was much discussed in the 1980s and 1990s but might well have released its energy in a so-called slow or "silent" earthquake in which pressures are released over hours or even months and are not necessarily detectable as the kind of motion we generally identify as an earthquake. Herb Dragert, Kelin Wang, and Thomas S. James, "A Silent Slip Event on the Deeper Cascadia Subduction Interface," *Science* 292, no. 5521 (2001): 1525–1528. Further, the idea that seismic pressure builds up and must be regularly released (the basis of the "gap" theory) has since lost some credibility, as Hough (2010) described.

63. CIRES's Espinosa Aranda and a Japanese engineer named Nakamura seem to have arrived at this unique application as solutions to their particular problems independently in the late 1980s. See Yutaka Nakamura, "On the Urgent Earthquake Detection and Alarm System (UrEDAS)," in *Proceedings of the 9th World Conference on Earthquake Engineering VII* (1988), 673–678., Juan Manuel Espinosa-Aranda, Alejandro Jimenez, O. Contreras, Gerardo Ibarrola, and R Ortega, "Mexico City Seismic Alert System," in *Simposio Internacional Sobre Prevencion de Desastres Sismicos* (Mexico City: CENAPRED, 1992).

64. Mexico City's executive is often referred to as a "governor," largely because the powers of the office are more like those of the executive offices of states than like those of the executive offices of other cities.

65. Quoted from presentation at Avances y Retos en Sismología, Ingeniería y Gestión de Riesgos a 30 Años del Sismo de 1985 (Mexico City, September 17–19, 2015).

66. See Juan Manuel Espinosa-Aranda, Armando Cuellar, Armando Garcia, Gerardo Ibarrola, Roberto Islas, Samuel Maldonado, and F. H. Rodriguez, "Evolution of the Mexican Seismic Alert System (SASMEX)," *Seismological Research Letters* 80, no. 5 (2009): 694–709.

67. Fundación Javier Barros Sierra, *Seminario Aprovechamiento del Sistema de Alerta Sísmica*, 8.

68. Gerardo Rico, Víctor Ruiz, Topiltzin Ochoa, and Carlos Camacho, "Sismo de 6.5 grados; Un muerto en el DF," *La Jornada*, July 16, 1996; Mirna Servín, "Alerta Sísmica," *La Jornada*, June 1, 1998. Angel Bolaños, "Restringida, la señal de alerta; se activó parcialmente (primera plana)," *La Jornada*, June 16, 1999; "Temblor de 5.3 Richter sacude la ciudad; no hubo víctimas ni daños," *La Crónica de Hoy*, September 26, 2002.

69. Elia Arjonilla Cuenca, "Evaluación de la alerta sísmica para la ciudad de México desde una perspectiva sociológica: Resultados en poblaciones escolares con y sin alerta," paper presented at Conference on Early Warning Systems for Reduction of Natural Disasters (Chile, April 27–30, 1998).

70. These were recorded in news articles from the period. Roger Díaz de Cossío and Antonio Alonso Concheiro, "Desgracias: Olvido y recordatorio," *Este Pais*, May 1, 2002. Miguel Bárcena, "El fangoso suelo del DF amplifica los sismos," *Epoca*, September 26, 1994; Elisa Robledo, Carolina Ballesteros Niño, Noé Cruz Serrano, Rebeca Hernández Marín, Miguel Bárcena, and Ernesto Zavaleta, "Sigue abierta la herida que dejo el sismo del 85," September 26, 1994.

71. Sistema (Agregado 1-SO17) XXX Costas, *Servicio Universal de Noticias*, September 19, 2000.

72. Complaints regarding this theme can be found in popular news articles from the time, which also describe a brief suspension of service. "Servicio alerta sismica," *Servicio Universal de Noticias*, September 3, 2000; "Escuelas alarma pública," *Servicio Universal de Noticias*, October 13, 2001; Sergio Morales, "Critican diputados suspensión del SAS," *El Economista*, March 13, 2003.

73. It is worth noting that this is not the dense network that Japan boasts. While ubiquitous sensing can structure our imagination of environmental monitoring, this kind of minimal but highly strategized spread is the reality of much knowledge production.

74. "Ciudad de México se mantiene alerta 25 años después de su peor terremoto," *Agencia EFE*, September 18, 2010; Ariette Gutiérrez, "Delegación Cuauhtémoc contrará con su alerta sísmica," *El Sol de México*, April 9, 2010.

75. Norma Técnica Complementaria al Reglamento de la Ley de Protección Civil del Distrito Federal 2010 (Government of Mexico City).

76. AlertaDF and SkyAlert were the main drivers here. I've written about these partially integrated infrastructures elsewhere. See Elizabeth Reddy, "Crying 'Crying

Wolf': How Misfires and Mexican Engineering Expertise Are Made Meaningful," *Ethnos* (2019): 1–16.

77. The company Grillo began to do this but soon opted to develop its own sensory network.

78. An unpublished market research survey by company MDreieck presented to CIRES while I was visiting also showed evidence of general ignorance of SASMEX.

79. In a useful typology of warning in alerting developed by McBride et al, this might be considered a "Late Alert." See S. K. McBride, A. Bostrom, J. Sutton, R. M. de Groot, A. S. Baltay, B. Terbush, P. Bodin, P. M. Dixon, E. Holland, R. Arba, P. Laustsen, S. Liu, and M. Vinci, "Developing Post-Alert Messaging for ShakeAlert, the Earthquake Early Warning System for the West Coast of the United States of America," *International Journal of Disaster Risk Reduction* 50 (2020).

80. Because there had been an earthquake in southern Mexico earlier in the month, several states canceled plans to participate in the drill. See: "Megasimulacros were cancelled in the states of Mexico, Michoacán, Guerrero, Oaxaca and Puebla." See "Secretaria de Gobernación Boletín No. 304," accessed December 27, 2019, https://www.gob.mx/segob/prensa/suspenden-os-estado.

81. Reports in Mexican popular media include: Sara Hidalgo and Andrés Lajous, "11 segundos vitales: ¿por qué no sonó a tiempo la alerta sísmica el 19 de Septiembre?" *Animal Politico*, November 1, 2017, https://www.animalpolitico.com/2017/11/alerta-sismica-sensores-cdmx; Gabriela Romero Sánchez, "Un minuto de silencio y luego el simulacro, piden damnificados," *La Jornada*, August 31, 2018, https://www.jornada.com.mx/2018/08/31/capital/032n2cap.

82. Twelve seconds was the final official time given by CIRES, though early reports on how much of a warning the earthquake early warning system had given were contradictory—some saying twenty seconds and others none whatever. Reports suggest that the quake was detected and analyzed by a first SASMEX station at 1:14:57 and confirmed by another 13:15:04. This was reported in Hidalgo and Lajous, "11 segundos vitales."

83. In 2018, authorities reported trouble with 12 percent of the loudspeakers meant to sound the earthquake early warning; there is reason to understand that this problem existed before the 2017 quake, as the popular media outlet *Animal Politico* reported. "Autoridades de CDMX reconocen que cientos de altavoces no emitieron alerta sísmica por ser obsoletos," *Redacción Animal Politico*, June 29, 2018, https://www.animalpolitico.com/2018/07/altavoces-alerta-sismica-cdmx.

CHAPTER 2

1. See typology of problematic alerts in S. K. McBride, A. Bostrom, J. Sutton, R. M. de Groot, A. S. Baltay, B. Terbush, P. Bodin, P. M. Dixon, E. Holland, R. Arba, P. Laustsen,

S. Liu, and M. Vinci, "Developing Post-Alert Messaging for ShakeAlert, the Earthquake Early Warning System for the West Coast of the United States of America," *International Journal of Disaster Risk Reduction* 50 (2020).

2. Early warning experts Dennis Mileti and John H. Sorensen have recommended that members of the public take advantage of warnings in this way for years. Dennis Mileti, *Communication of Emergency Public Warnings: A Social Science Perspective and State-of-the-Art Assessment* (Washington, DC: Federal Emergency Management Agency, 1990); John H. Sorensen, "Hazard Warning Systems: Review of 20 Years of Progress," *Natural Hazards Review* 1, no. 2 (2000): 119–125. Drilling, rehearsals, or practices of this sort can have political as well as practical effects. See Tracy C. Davis, *Stages of Emergency: Cold War Nuclear Civil Defense* (Durham, NC: Duke University Press, 2007); Ben Anderson and Peter Adey, "Affect and Security: Exercising Emergency in 'UK Civil Contingencies,'" *Environment and Planning D: Society and Space* 29, no. 6 (2011): 1092–1109.

3. See S. K. McBride, A. Bostrom, J. Sutton, R. M. de Groot, A. S. Baltay, B. Terbush, P. Bodin, P. M. Dixon, E. Holland, R. Arba, P. Laustsen, S. Liu, and M. Vinci, "Developing Post-Alert Messaging for ShakeAlert, the Earthquake Early Warning System for the West Coast of the United States of America," *International Journal of Disaster Risk Reduction* 50 (2020).

4. The amount of information that can be communicated by sirens makes this more of an "alert" than a "warning," as the latter can be understood to contain more information. I use "warning" here because it is in keeping with accepted language about what this type of system does. See Jeanette Sutton, Brett Hansard, and Paul Hewett, "Changing Channels: Communicating Tsunami Warning Information in Hawaii," *Proceedings of the 3rd International Joint Topical Meeting on Emergency Preparedness and Response, Robotics, and Remote Systems* (2011): 1–14. I do not parse this difference here because I did not observe the distinction used among either residents of Mexican City or the engineers at CIRES. I note it for curious readers.

5. "One moment" are the words of Dr. Sergio Puente Aguilar, voiced in an interview I conducted in 2014. Puente is a researcher and professor at the Centro de Estudios Demográficos, Urbanos y Ambientales at El Colegio de Mexico, and the author of a number of works on risk in urban Mexico. See Sergio Puente Aguilar, "Social Vulnerability to Disasters in Mexico City: An Assessment," in *Crucibles of Hazard: Mega-Cities and Disasters in Transition*, ed. J. K. Mitchell (Tokyo: The United Nations University Press, 1999); and "Un megalopolis en riesgo: La ciudad de Mexico," in *Los Grandes Problemas de Mexico: Medio Ambiente*, ed. B. Graizbord and J. L. Lezama (Mexico City: El Colegio de Mexico, 2013). He was also on the scientific advisory committee of CIRES, and though he hosted long meetings about SASMEX, he considered earthquake early warning to be one mode among many for intervening on earthquake risk management.

6. With this term I reference two movements in science and technology studies. First, I point to work like that of sociologist John Law that accounts for technological systems in ways that take social and technical forces, logics, and agents seriously, as in "Technology, Closure and Heterogeneous Engineering: The Case of the Portuguese Expansion," in *The Social Construction of Technological Systems*, ed. W. Bijker, T. Hughes, and T. Pinch (Cambridge, MA: MIT Press, 1987): 111–134. Second, I also reference efforts like anthropologist Helen Verran's that address forces, ideas, and conditions more broadly as "heterogenous material-symbolic assemblages" that, as she writes, are active in "making and remaking our world." Helen Verran, "Re-Imagining Land Ownership in Australia," *Postcolonial Studies: Culture, Politics, Economy* 1, no. 2 (1998): 250. I use this word, in short, to signal my attention to multiplicity within and around this system.

7. I use the term "embeddedness" to resonate with an analogous scholarly conversation around formal economic systems and their sociocultural contexts. See Karl Polanyi, *The Great Transformation: The Political and Economic Origins of Our Time* (New York: Farrar & Rinehart, 1944); Clifford Geertz, *Peddlers and Princes: Social Change and Economic Modernization in Two Indonesian Towns* (Chicago: University of Chicago Press, 1963); Mark Granovetter, "Economic Action and Social Structure: The Problem of Embeddedness," *American Journal of Sociology* 91 (1985): 481–510. In much the same way that scholars have engaged with the challenge of how best to understand convergences and contradictions related to the global rise of the market economy, the question of how an early warning can (or cannot) be made part of ordinary life is a site of ongoing trouble for emergency managers and technoscientists.

8. Craig Calhoun, "The Idea of Emergency: Humanitarian Action and Global (Dis) Order," in *Contemporary States of Emergency: The Politics of Military and Humanitarian Interventions*, ed. D. Fassin and M. Pandolfi (New York: Zone Books, 2010), 30. It is worth making clear that this is a critical social scientific analysis of emergency that I am using to help me historicize and reflect on the category of emergency. Professionals in fields related to disaster response and recovery also draw on a sizeable body of research literature, or *literatures*, related to risk reduction, preparedness for, and recovery from emergencies. See George Haddow, Jane Bullock, and Damon P. Coppola, *Introduction to Emergency Management* (Newton, MA: Butterworth-Heinemann, 2013).

9. As Sara McBride shows, while emergency management and emergency communication are related in important and practical ways, the research that informs these practices has significant divergences as well as relations. Although communications research often highlights complex emergent relations, she notes that many "current civil defence public education campaigns still maintain . . . the 'tell people what to do and they will do it' ideology." Sara K. McBride, *The Canterbury Tales: An Insider's Lessons and Reflections from the Canterbury Earthquake Sequence to Inform Better Communication Models* (PhD diss., Massey University, 2017), 65.

10. Mileti and Sorensen, *Communication of Emergency Public Warnings*, 2–10. These warnings are, at times, more directive [see Ann Bostrom, Cynthia J. Atman, Baruch Fischhoff, and M. Granger Morgan, "Evaluating Risk Communications: Completing and Correcting Mental Models of Hazardous Processes, Part II," *Risk Analysis* 14, no. 5 (1994): 789–798] and at others, more processual, as in Julia Becker, Sally H. Potter, Sara K. McBride, Anne Wein, E. E. H. Doyle, and D. Paton, "When the Earth Doesn't Stop Shaking: How Experiences over Time Influenced Information Needs, Communication, and Interpretation of Aftershock Information during the Canterbury Earthquake Sequence, New Zealand," *International Journal of Disaster Risk Reduction* 34 (2019): 397–411.

11. See Jeannette Sutton, Sarah C. Vos, Michele M. Wood, and Monique Turner, "Designing Effective Tsunami Messages: Examining the Role of Short Messages and Fear in Warning Response," *Weather, Climate, and Society* 10, no. 1 (2018): 75–87; Hamilton Bean, Brooke F. Liu, Stephanie Madden, Jeannette Sutton, Michele M. Wood, and Dennis S. Mileti, "Disaster Warnings in Your Pocket: How Audiences Interpret Mobile Alerts for an Unfamiliar Hazard," *Journal of Contingencies and Crisis Management* 24, no. 3 (2016): 136–147.

12. On the use of social media and other short-form modes of emergency communication, see Jeannette Sutton, E. S. Spiro, B. Johnson, S. Fitzhugh, B. Gibson, and C. T. Butts, "Warning Tweets: Serial Transmission of Messages during the Warning Phase of a Disaster Event," *Information, Communication & Society* 17, no. 6 (2014): 765–787; and Jeannette Sutton and Erica D. Kuligowski, "Alerts and Warnings on Short Messaging Channels: Guidance from an Expert Panel Process," *Natural Hazards Review* 20, no. 2 (2019).

13. There is a substantial body of literature that parses these understandings and actions. They are reviewed admirably in Ann Bostrom, Adam L. Hayes, and Katherine M. Crosman, "Efficacy, Action, and Support for Reducing Climate Change Risks," *Risk Analysis* 39, no. 4 (2019): 805–828.

14. While I note that personal orientation toward surprises may matter, I want to be clear: there is consensus among researchers that stories about widespread public panic upon receiving warnings, much like those about post-disaster lawlessness and shock responses, are largely myths. For a serious discussion of how ideas about panic have surfaced and circulated, see Scott Knowles, *The Disaster Experts: Mastering Risk in Modern America* (Philadelphia: University of Pennsylvania Press, 2012).

15. Although they may not be ideal, our slow responses may be far from an unusual way to interact with this technology. Two recent surveys of Japanese earthquake early warning users found that many respondents simply used the warning to mentally brace themselves and took no significant physical actions. See Kazuya Nakayachi, Julia S. Becker, Sally H. Potter, and Maximilian Dixon, "Residents' Reactions to Earthquake Early Warnings in Japan," *Risk Analysis* 39, no. 8 (2019).

16. On the tension between descriptive and brief messages, see Michele M. Wood, Dennis S. Mileti, Hamilton Bean, Brooke F. Liu, Jeannette Sutton, and Stephanie Madde, "Milling and Public Warnings," *Environment and Behavior* 50, no. 5 (2018): 535–566.

17. Sara K. McBride, Anne Bostrom, Jeanette Sutton, Robert M. de Groot, A. S. Baltay, Brian Terbush, Paul Bodin, Maximilian Dixon, E. Holland, R. Arba, P. Laustsen, S. Liu, and M. Vinci, "Developing Post-Alert Messaging for ShakeAlert, the Earthquake Early Warning System for the West Coast of the United States of America," *International Journal of Disaster Risk Reduction* 50 (2020): 101713; Jeanette Sutton, Laura Fischer, Lori E. James, and Sarah E. Sheff, "Earthquake Early Warning Message Testing: Visual Attention, Behavioral Responses, and Message Perceptions," *International Journal of Disaster Risk Reduction* 49 (2020): 101664. Jeanette Sutton and Erica D. Kuligowski, "Alerts and Warnings on Short Messaging Channels: Guidance from an Expert Panel Process," *Natural Hazards Review* 20, no. 2 (2019): 04019002.

18. Quote from interview. Translation by author.

19. This term, which refers to the practice of characterizing "micro zones" of soil within geographic area according to their response to seismicity, is described in greater detail in chapter 1.

20. That is not to say that these distinctions do not still matter when Mexico City is only a fraction of the Mexico City Metro Area, and when regional Civil Protection agencies are clients who choose to pay for earthquake early warning services or not.

21. CONACYT-INEGI, *Encuesta sobre la percepción pública de la ciencia y la tecnología (ENPECYT)* (Mexico City: INEGI, 2014).

22. CIRES engineers continued to maintain SASPERs, but they hoped schools would begin to adopt less expensive emergency radios instead.

23. The Spanish word they use, "gabinete," can also refer to the tower that holds the CPU of a desktop computer. This means that in Spanish, the word has a stronger association with digital technology than it does when it is translated to English.

24. In this case, as in a great deal of emergency management practice, "education" refers to an activity by which experts inform publics about hazards before they happen. See Dennis Mileti, S. Nathe, P. Gori, M. Greene, and E. Lemersal, *Public Hazards Communication and Education: The State of the Art* (Boulder, CO: Natural Hazards Center University of Colorado, 2004). This type of "education" may rely, as McBride notes, on a model more concerned with transferring information to learners than supporting their acquisition of new critical skills. See *The Canterbury Tales*, 60.

25. The outreach team comprised primarily women. While most CIRES departments are mixed-gender, women are usually in the minority in technical spaces. This was an exception.

26. Norma Técnica Complementaria al Reglamento de la Ley de Protección Civil del Distrito Federal, 2010, Mexico City's Law of Civil Protection.

27. Gerardo Suárez, David Novelo, and Elizabeth Mansilla, "Performance Evaluation of the Seismic Alert and a Social Perspective," *Seismological Research Letters* 80, no. 5 (2009): 707–716.

28. Revet notes that children are often key to efforts to illustrate disasters. She notes the symbolic power of their presence and writes that they "play a role that fluctuates artfully between innocence, vulnerability, and resilience." Sandrine Revet, *Disasterland: An Ethnography of the International Disaster Community* (Paris: The Sciences Po Series in International Relations and Political Economy, 2020), 59.

29. It should be noted that communicating in terms of magnitude at source is a debated practice. While broadly used by publics, experts find that the measure does not communicate the key concern of risk management, which is less about the magnitude at source than the intensity of shaking in communities at risk. For that reason, CIRES engineers use magnitude as a resource for public communication related to the system only sometimes, when it is useful. Like many earthquake experts, they prefer to avoid doing so. In this way, magnitude language is only one way to talk earthquakes, a resource to be used when necessary. See chapter 1 for a more in-depth explanation of how the system is arranged in Mexican territory and how it came to be so.

30. Given a multiple-choice question about expectations, 58.3 percent of 2,400 people in Mexico City and the surrounding region who were surveyed indicated that they expected sixty seconds of warning time. Only 22.8 percent indicated the correct answer: the time varies. On a 5-point Likert scale, 77.5 percent of respondents considered themselves knowledgeable or very knowledgeable about response—but the study did not inquire about what response they thought would be appropriate. See Jaime Santos-Reyes, "How Useful Are Earthquake Early Warnings? The Case of the 2017 Earthquakes in Mexico City," *International Journal of Disaster Risk Reduction*, 2019. Earthquake early warnings can be hard to parse, though; see surveys on Japanese understanding of earthquake early warning systems reviewed in James D. Goltz and Evelyn Roeloffs, "Imminent Warning Communication: Earthquake Early Warning and Short-Term Forecasting in Japan and the US James," in *Disaster Risk Communication and Communities: A Challenge from a Social Psychological Perspective*, ed. Katsuya Yamori (Singapore: Springer, 2020).

31. For an earthquake originating in Oaxaca on September 7, 2017, when the Mexico City region had significant warning time, respondents were asked to rank how useful they thought the warning was on a five-point Likert scale. Only 50.2 percent of respondents indicated they found it useful or very useful. For an earthquake originating in Puebla on September 19 (discussed in chapter 1), only 14.3 percent indicated that they found it useful or very useful. See Santos-Reyes, "How Useful Are Earthquake Early Warnings?"

32. Use of SASMEX in schools was never universal, even in areas that the system nominally protected. Nevertheless, CIRES rallied attention around their work in schools.

33. Direct quote from interview. Translated by the author.

34. Direct quote from interview. Translated by the author.

35. Vivian G. Gratton, Herbert D. Thier, Elia Arjonilla, and Rosa Melgar, "The Recovery of Schools from Earthquake Effects: Lessons from Mexico City," *Disasters* 11, no. 4 (1987): 310–316.

36. "Sólida planeación y preparación por parte de la comunidad." See Elia Arjonilla, "Evaluación de la alerta sísmica para la ciudad de México desde una perspectiva sociológica: Resultados en poblaciones escolares con y sin alerta," *International IDNDR: Conference on Early Warning Systems for Reduction of Natural Disasters*, 1998, 3.

37. Arjonilla, "Evaluación de la alerta sísmica," 2–3.

38. Suárez, Novelo, and Mansilla, "Performance Evaluation."

39. The business that supplied these, MDreieck, was founded by former CIRES engineers.

40. Data on radios distributed received from CIRES internal document: "Historico de instalaciones de equipos receptores por parte del CIRES," accessed August 13, 2014. On trouble with distribution, see: Paris Martínez, "El gobierno compra alertas sísmicas, pero éstas desaparecen o sequedan en bodegas," *Animal Politico*, September 12, 2017, http://www.animalpolitico.com/2017/09/gobierno-compraalertas-sismicas -desaparecen-bodegas/.

41. Direccion de Emergencia Escolar, *Informacion Relativa al Sistema de Alerta Sismica* (Mexico City: Secretaría de Educación Pública, 1995), from CIRES collection.

42. Direct quotes from conversation in public space. Translated by the author.

43. Direct quotes from Twitter. Translated by the author.

44. That is, "for the fear," the implication being that they wanted to eat comforting food to calm themselves after the shock.

45. As reported in Arturo Páramo, "Sismo de 4.6 grados a medianoche alertó al D. F.," *Excelsior*, September 30, 2015, http://www.excelsior.com.mx/comunidad/2015 /09/30/1048507#view-1.

46. Arthur J. Rubel is well known for his extensive work exploring maladies related to *susto* or "fright" in Latin America. See A. J. Rubel, "Concepts of Disease in Mexican-American Culture," *American Anthropologist* 62, no. 5 (1960): 795–814; and A. J. Rubel, "The Epidemiology of a Folk Illness: Susto in Hispanic America," *Ethnology* 3, no. 3 (1964): 268–283. For a more contemporary overview, see S. C. Weller, R. D. Baer, J. Garcia de Alba Garcia, M. Glazer, R. Trotter, L. Pachter, and R. E. Klein, "Regional Variation in Latino Descriptions of *Susto*," *Culture, Medicine and Psychiatry* 26 (2002): 449–472.

47. Direct quotes from Twitter. Translated by the author.

48. I have written about "crying wolf" as a way for people to voice concern about early warning elsewhere. See Elizabeth Reddy, "Crying 'Crying Wolf' How Misfires and Mexican Engineering Expertise Are Made Meaningful," *Ethnos* 85, no. 2 (2020): 335–350.

49. Priscila Navarrete, "Alerta sísmica en México por dos temblores en 12 horas," *El País*, September 30, 2015, http://internacional.elpais.com/internacional/2015/09/30 /actualidad/1443635879_427581.html.

50. See K. Landeros-Mugica, J. Urbina-Soria, and I. Alcántara-Ayala, "The Good, the Bad and the Ugly: On the Interactions among Experience, Exposure and Commitment with Reference to Landslide Risk Perception in México," *Natural Hazards* 80, no. 3 (2016): 1515–1537.

51. Arturo Páramo, "Alerta Sísmica busca advertir, no asustar," *Excelsior*, October 1, 2015, https://www.excelsior.com.mx/comunidad/2015/10/01/1048733.

52. A magnitude 4.9 quake, which happened at 6:29 p.m.

53. Direct quote from conversation. Reprinted with permission.

CHAPTER 3

1. Richard Allen, "Welcome and Opening Remarks," *Third International Conference on Earthquake Early Warning*, September 3, 2014.

2. Japan, China, Italy, Switzerland, Mexico, the Caribbean, Canada, and, of course, the United States were all very well represented.

3. Attendees included future California secretary of state and US senator Alex Padilla, at the time a state senator; future California governor Gavin Newsom, then serving as lieutenant governor; and San Francisco mayor Ed Lee.

4. Japan's system is the only exception. Although a number of articles about how earthquake early warning were published by Californian scientists in the late 1980s, their early experimental system had a long way to go before it became the basis for ShakeAlert. See T. H. Heaton, "A Model for a Seismic Computerized Alert Network," *Science* 228 (1985): 987–990; and R. Holden, M. Reichle, and R. Lee, *Technical and Economic Feasibility of an Earthquake Warning System in California* (Special Publication 101: California Division of Mines and Geology 1989). A longer history is detailed in James D. Goltz and Evelyn Roeloffs, "Imminent Warning Communication: Earthquake Early Warning and Short-Term Forecasting in Japan and the US," in *Disaster Risk Communication and Communities: A Challenge from a Social Psychological Perspective*, ed. Katsuya Yamori (Singapore: Springer, 2020).

5. The observation that early warning systems are often designed to meet technological requirements and may neglect issues related to use emerges from my research

and has been noted in related studies. See Jeannette Sutton and Erica D. Kuligowski, "Alerts and Warnings on Short Messaging Channels: Guidance from an Expert Panel Process," *Natural Hazards Review* 20, no. 2 (2019): 1–10.

6. Anthropogenic earth motions are also, technically, earthquakes. When stadiums full of sports fans make the ground shake, it is not necessarily interesting from a seismological or emergency management perspective. However, the same cannot be said of other kinds of seismicity that humans produce. Emerging research suggests, for instance, that the seismic effects of hydraulic fracturing practices may take on importance for both in the coming years. See Elizabeth Reddy, "Stability," in *Anthropocene Unseen: A Lexicon*, ed. C. Howe and A. Pandian (Goleta, CA: Punctum Press, 2019). Seismological knowledge has also been crucial to facilitating and documenting global nuclear arms testing. See Kai Henrik Barth, "The Politics of Seismology: Nuclear Testing, Arms Control, and the Transformation of a Discipline," *Social Studies of Science* 33, no. 5 (2003): 743–781. Indeed, military funding for this research has had powerful effects for the twentieth-century disciplinary formation of geophysics, as described in Ronald E. Doel, "Constituting the Postwar Earth Sciences: The Military's Influence on the Environmental Sciences in the USA after 1945," *Social Studies of Science* 33, no. 5 (2003): 635–666. It has for scientific modes of understanding earth systems more broadly, too, as in Joseph Masco, "Bad Weather: On Planetary Crisis," *Social Studies of Science* 40, no. 1 (2010): 7–40.

7. This mostly happens at or near the edges of tectonic plates—hence the "Ring of Fire" around the Pacific Ocean. Intraplate faulting systems are also significant, though: these produced the New Madrid earthquakes in 1811 and 1812, the significance of which for political and scientific knowledge has been well documented by Conevery Bolton Valencius, *The Lost History of the New Madrid Earthquakes* (Chicago: University of Chicago Press, 2013). Another intraplate rupture produced the 2017 Puebla earthquake discussed in chapter 1 of this book.

8. For excellent detailed and accessible overviews of contemporary mainstream scientific ways of understanding seismicity, see Susan E. Hough, *Earthshaking Science: What We Know (and Don't Know) about Earthquakes* (Princeton, NJ: Princeton University Press, 2002).

9. A careful overview of contemporary work in the histories of scientific knowledge related to seismicity far beyond the scope of this chapter, but for a more thorough engagement, see Deborah R. Coen, *The Earthquake Observers: Disaster Science from Lisbon to Richter* (Chicago: University of Chicago Press, 2013). Seismological research in China and Japan was ongoing and represented less significant breaks with folk traditions than those cultivated in Europe. While Mexico was not a site of innovation, it was nonetheless an important location for seismic instrumentation; see Fa-ti Fan, "'Collective Monitoring, Collective Defense': Science, Earthquakes, and Politics in Communist China." *Science in Context* 25, no. 1 (2007): 127–154; and G. K. Clancey, *Earthquake Nation: The Cultural Politics of Japanese Seismicity, 1868–1930*

(Berkeley: University of California Press, 2006). Mexico's first seismographs—massive Wiechert devices weighing between twelve and seventeen tons each—were installed just to the southwest of Mexico City's historic center in 1910, only thirteen years after the first seismograph in North America was positioned in San Jose, California, as Roberto Quaas Weppen describes in "Breve Historia de La Instrumentacion Sismica En Mexico," in *Instrumentación sísmica de temblores fuertes* (Mexico City: CENAPRED and JICA, 1992), 35–40. At that time, this kind of state investment in science was explicitly political: "algo para daría lustre al país y le permitería superar el atraso, para formar parte de los países modernos," that is, "Something that to give luster to the country and enable it to overcome its backwardness and gain a place among the modern countries of the world." Natalia Priego, *Ciencia, historia y modernidad: La microbiología en México durante el porfiriato* (Madrid: Editorial CSIC-CSIC Press, 2009), 21.

10. Magnitude is often referred to as the "Richter scale," in reference to Charles Richter. California seismologist Richter and his colleague Beno Gutenberg developed a logarithmic method for describing the magnitude of seismicity in the 1930s based on the kinds of data that a particular type of seismometer could produce. This type of seismometer was not, as later researchers discovered, as useful for registering bigger earthquakes as it was for smaller ones, and consequently a new scale called "moment magnitude" has since come into more popular use (see Coen, *The Earthquake Observers*). On Richter's life and twentieth-century seismology, see Susan Hough, *Richter's Scale: Measure of an Earthquake, Measure of a Man* (Princeton, NJ: Princeton University Press, 2007).

11. See, for example, Virginia Garcia Acosta, Rocío Hernández, Irene Márquez, América Molina, Juan Manuel Pérez, Teresa Rojas, and Cristina Sacristán, "Cronologia de los sismos en la Cuenca del Valle de Mexico," in *Estudios sobre sismicidad en el Valle de Mexico*, ed. Sergio Alcocer (Mexico City: Departamento Del Distrito Federal Secretaria General de Obras, 1988) on the slow incorporation of time into eighteenth-century colonial seismic reports as public clocks appeared and in relation to concerns that earthquakes might be an indication of the degree of a Catholic god's anger; Coen, *The Earthquake Observers* on how expertise and power were at play in the development and policing of earthquake knowledge between the eighteenth and twentieth centuries; Valencius, *The Lost History of the New Madrid Earthquakes* on how evidence of earthquakes in the central United States has been used in various political and scientific projects; and Finn, *Documenting Aftermath* on how communication in crisis shapes how we understand events like earthquakes. See also Megan Finn, "Information Infrastructure and Descriptions of the 1857 Fort Tejon Earthquake," *Information and Culture: A Journal of History* 48, no. 2 (2013): 194–221.

12. For a more detailed historical account of these movements, see Sandrine Revet, *Disasterland: An Ethnography of the International Disaster Community* (Paris: The Sciences Po Series in International Relations and Political Economy, 2020), 164–168.

13. This does not include tsunamis. See Centre for Research on the Epidemiology of Disasters and UN Office for Disaster Risk Reduction, "Poverty & Death: Disaster Mortality 1996–2015" (2016), 1–20.

14. Maura R. O'Connor, "Two Years Later, Haitian Earthquake Death Toll in Dispute," *Colombia Journalism Review*, January 12, 2012, https://archives.cjr.org/behind_the_news/one_year_later_haitian_earthqu.php. The Interamerican Development Bank estimated it could cost as much as $13.9 billion to rebuild Haiti over time, a fraction of the impact to New Zealand. Eduardo Cavallo, Andrew Powell, and Oscar Becerra, "Estimating the Direct Economic Damage of the Earthquake in Haiti," IDB Working Paper No. IDB-WP-163 (2010). Real impacts are notoriously hard to assess, however.

15. The cost of this event, coupled with a second earthquake the next year, totaled $28 billion USD—the equivalent of almost 20 percent of the country's current gross domestic product. See Canterbury Earthquake Recovery Authority, "Funding the Recovery: The CERA Perspective," *EQ Recovery Learning*, April 18, 2016.

16. Port-au-Prince's population was 900,000 in 2010, a little less than three times Christchurch's 377,000. The difference in fatalities is not even close to proportional.

17. Mary Comerio, "Disaster Recovery and Community Renewal: Housing Approaches," *Cityscape: A Journal of Policy Development and Research* 16, no. 2 (2014): 51–68.

18. See overviews in Revet, *Disasterland*; Malcom Spector and John I. Kitsuse, *Constructing Social Problems* (New York: Routledge, 1987); Anthony Oliver-Smith, "Anthropological Research on Hazards and Disasters," *Annual Review of Anthropology* 25 (1996); Anthony Oliver-Smith and Susannah Hoffman, eds., *Catastrophe and Culture: The Anthropology of Disaster* (Santa Fe: School of American Research Press, 2002); Anthony Oliver-Smith and Susannah Hoffman, eds., *The Angry Earth: Disaster in Anthropological Perspective* (New York: Routledge, 1999); Kathleen J. Tierney "From the Margins to the Mainstream? Disaster Research at the Crossroads," *Annual Review of Sociology* 33, (2007): 503–525. A. J. Faas and Roberto Barrios, "Applied Anthropology of Risks, Hazards, and Disasters," *Human Organization* 74, no. 4 (2015): 287–295. This research has particularly benefited from the insights of cultural and political ecologists. See Aletta Biersack, "Reimagining Political Ecology: Culture/Power/History/Nature," in *Reimagining Political Ecology*, ed. Aletta Biersack and James B. Greenberg (Durham, NC: Duke University Press, 2006), 16–53.

19. See James Kendra and Joanne Nigg, "Engineering and the Social Sciences: Historical Evolution of Interdisciplinary Approaches to Hazard and Disaster," *Engineering Studies* 6, no. 3 (2014): 134–58. This has happened alongside what Kathleen Tierney has called the "mainstreaming" of attention to disaster in social sciences in her article "From the Margins to the Mainstream? Disaster Research at the Crossroads."

20. As UN Disaster policy official Andrew Maskrey put it in 2015, disaster is "endogenous" to the material and social organization of life. Maskrey went so far as to title an influential book *Los disastres no son naturales* [Disasters Are Not Natural] (1993) to emphasize the degree to which human decisions, not earthquakes themselves, make the destruction of the built environment and loss of life possible. Maskrey notes what he calls "a sea change" in the way disaster prevention was conceptualized and done throughout the late-twentieth century. See Andrew Maskrey, "Gestion de

riesgos: Cuenta historico," paper presented at Sismo 85 Conference (Mexico City, September 18, 2015); and *Los disastres no son naturales* (Panama City, PN: Red de Estudios Sociales en Prevención de Desastres en América Latina, 1993). See also Scott Knowles, *The Disaster Experts: Mastering Risk in Modern America* (Philadelphia: University of Pennsylvania Press, 2013); and Joseph Masco, *The Theater of Operations: National Security Affect from the Cold War to the War on Terror* (Durham, NC: Duke University Press, 2014) on simultaneous shifts in risk and security logics.

21. In Mexico, such insights are written into state-sponsored research and emergency management institutions such as CENAPRED, the National Center for Disaster Prevention.

22. A correspondence between Rousseau and Voltaire is often cited as the first evidence of a "modern" perspective on human agency in the production of disasters. Russell R. Dynes, "The Dialogue between Voltaire and Rousseau on the Lisbon Earthquake: The Emergence of a Social Science View," *International Journal of Mass Emergencies and Disasters* 18 (2000): 97–115.

23. Wescoat offers an expanded genealogy of some of these concepts in policy and academic research in James L. Wescoat Jr., "Political Ecology of Risk, Hazards, Vulnerability, and Capacities," in *The Routledge Handbook of Political Ecology*, ed. Tom Perreault, Gavin Bridge, and James McCarthy (London: Routledge, 2015), 293–302.

24. The sociologist and educator Elia Arjonilla's short book on this topic, originally written for Mexican educators to help them communicate with their students, articulates the utility of this heuristic nicely: "En un mundo que ya no se basa en certezas tradicionales, lo más riesosos . . . puede ser que no sean capaces de tomar decisions"; that is, "In a world no longer based on traditional certainties, the riskiest thing . . . could be not having the capacity to make decisions." *Como hablar de riesgo: Consideraciones teóricas y unidas, temáticas con ejercitos para la escuela* (Mexico City: Fundación Mexicana para la Salud, n.d.), 10.

25. This kind of exercise in defining variables and relationships is not limited to disaster professionals in fields related to disaster response and recovery. Andrea Ballestero considers how a similar formula enfolds issues of morality in flexible decision-making. See Andrea Ballestero, *A Future History of Water* (Durham, NC: Duke University Press, 2019), 37–74

26. The Office of the United Nations Disaster Relief Coordinator, "Natural Disasters and Vulnerability Analysis." *Report of Expert Group Meeting* (Geneva: United Nations, 1979), iv and 6.

27. "Natural Disasters and Vulnerability Analysis," 5.

28. "Natural Disasters and Vulnerability Analysis," 5. In Mexico, I have not seen this operationalization taken up significantly as more than a useful way of illustrating the relationship proposed by the equation itself.

29. Some have called this a "physicalist" approach to mitigating risk. See Christine Gibb, "A Critical Analysis of Vulnerability," *International Journal of Disaster Risk Reduction* 28 (2018): 327–334; and Kathleen Tierney, "From the Margins to the Mainstream? Disaster Research at the Crossroads."

30. The shift was hardly universal, and, when he wrote his influential book in the early 1990s, Andrew Maskrey described approaches to vulnerability characterized by what he called a spectrum of approaches, which included as its poles the idea that "disasters were characteristic of natural hazards" and the idea that they might be a matter of "socioeconomic and political structures and practices," concerning "construction and settlement patterns" somewhere the middle of the two. See Andrew Maskrey, *Los desastres no son naturales*, 2.

31. The recent UN-sponsored Sendai Framework on Disaster Risk Reduction (2015) addresses social and structural issues as part of a single problem: systemically and structurally vulnerable people are disproportionately affected by disaster. This is a significant change even from the focus of the 2005 Hyogo framework that predated it, which focused more on disasters themselves rather than the factors that contribute to them. See United Nations International Strategy for Disaster Reduction, *Hyogo Framework for Action 2005–2015: Building the Resilience of Nations and Communities to Disasters* (Hyogo, Japan: United Nations, January 22, 2005); and United Nations International Strategy for Disaster Reduction, *Sendai Framework for Disaster Risk Reduction 2015–2030* (Sendai, Japan: United Nations. March 18, 2015).

32. Anthony Oliver-Smith and Susanna Hoffman write that "disaster becomes unavoidable in the context of a historically produced pattern of 'vulnerability'" (*Catastrophe and Culture*, 3). However, as historian Gregory Clancey reminds us, we would do well to beware of reducing natural events to social ones. See Gregory Clancey, "The Meiji Earthquake: Nature, Nation, and the Ambiguities of Catastrophe," *Modern Asian Studies* 40, no. 4 (2006): 818.

33. Though ideas about physical vulnerability due to urbanization continue to motivate disaster studies scholarship, Tierney notes that the vast majority of new university programs on risk and disaster are housed in public administration, engineering, geography, and urban planning units, rather than in social science departments ("From the Margins to the Mainstream? Disaster Research at the Crossroads," 517).

34. For example, Sabine Loos, David Lallemant, Feroz Khan, Jamie McCaughey, Robert Banick, Nama Budhathoki, and Jack Baker, "Beyond Building Damage: Estimating and understanding non-recovery following disasters," *Preprint through Research Square* (2021).

35. See Roberto E. Barrios, "Resilience: A Commentary from the Vantage Point of Anthropology," *Annals of Anthropological Practice* 40, no. 1 (2016): 28–38; Jeremy Walker and Melinda Cooper, "Genealogies of Resilience from Systems Ecology to the Political Economy of Crisis Adaptation," *Security Dialogue* 42, no. 2 (2011): 143–160; J. C. Gaillard, "Vulnerability, Capacity, and Resilience: Perspectives for Climate and Development Policy," *Journal of International Development* 22 (2010); and John

Hausdoerffer, "What Anything Is For," in *Pragmatist and American Philosophical Perspectives on Resilience*, ed. Kelly A. Parker and Heather E. Keith (Lanham, MD: Lexington Books, 2019).

36. Indeed, the UN Office for Disaster Risk Reduction has recommended early warning systems particularly for places subject to such conditions, suggesting that they are "far more cost-effective in strengthening coping mechanisms than is primary reliance on post-disaster recovery" (*Hyogo Framework*, 5).

37. Other kinds of warning of interest might include those documented in C. Garcia and C. J. Fearnley, "Evaluating Critical Links in Early Warning Systems for Natural Hazards," *Environmental Hazards* 11, no. 2 (2012): 123–137; J. H. Sorensen, "Hazard Warning Systems: Review of 20 Years of Progress," *Natural Hazards Review* 1, no. 2 (2000): 119–125; J. S. Becker, G. S. Leonard, S. H. Potter, M. A. Coomer, D. Paton, K. C. Wright, and D. M. Johnston, "Organisational Response to the 2007 Ruapehu Crater Lake Dam-Break Lahar in New Zealand: Use of Communication in Creating an Effective Response," in *Observing the Volcano World. Advances in Volcanology*, ed. C. J. Fearnley, D. K. Bird, K. Haynes, W. J. McGuire, and G. Jolly (Barcelona, Spain: Springer, 2017); J. Brotzge and W. Donner, "The Tornado Warning Process: A Review of Current Research, Challenges, and Opportunities," *Bulletin of the American Meteorological Society* 94, no. 11 (2013): 1715–1733.

38. On the ways that this community distinguishes between forecasting and prediction, see Revet, *Disasterland*. Some probabilistic long- and short-term forecasting is possible, however; see James D. Goltz and Evelyn Roeloffs, "Imminent Warning Communication: Earthquake Early Warning and Short-Term Forecasting in Japan and the US," in *Disaster Risk Communication and Communities: A Challenge from a Social Psychological Perspective*, ed. Katsuya Yamori (Singapore: Springer, 2020); and E. Tapia-Hernández, E. Reddy, and L. J. Oros-Aviles, "Earthquake Predictions and Scientific Forecast: Dangers and Opportunities for a Technical and Anthropological Perspective," *Earth Sciences Research Journal* 23, no. 4 (2019).

39. Ignacio Farías, for example, writes about the troubles of recognizing such hazards and the obligation to do so. See "Misrecognizing Tsunamis: Ontological Politics and Cosmopolitical Challenges in Early Warning Systems," in *Disasters and Politics: Materials, Experiments, Preparedness*, 62 (2014): 61–87. In her book on weather prediction, Phaedra Daipha suggests that, while the way experts grapple with uncertainty can be concealed in decision-making, it nevertheless remains essential to understand. See *Masters of Uncertainty: Weather Forecasters and the Quest for Ground Truth* (Chicago: University of Chicago Press, 2015).

40. On the state of prediction in earthquake science, see Susan Hough, *Predicting the Unpredictable: The Tumultuous Science of Earthquake Prediction.* (Princeton, NJ: Princeton University Press, 2010).

41. This nineteenth-century plan was even used to help settle a legal battle in Mexico. A group of geophysicists patented the idea of an early alert system in 1986.

In 1993, they brought a legal suit against both the government of Mexico City and the nonprofit through which the earthquake early warning was being run, asking for payment of damages and fees to the tune of $20 million pesos (the equivalent of a little less than $6 million USD). The suit lasted nearly a decade. The patent was finally nullified in 1999, in light of pre-existing work on earthquake early warning.

42. Richard M. Allen, Paolo Gasparini, O. Kamigaichi, and M. Bose, "The Status of Earthquake Early Warning around the World: An Introductory Overview," *Seismological Research Letters* 80, no. 5 (2009): 682–693; W. H. K. Lee and J. M. Espinosa-Aranda, "Earthquake Early-Warning Systems: Current Status and Perspectives," in *Early Warning Systems for Natural Disaster Reduction*, edited by J. Zschau and A. N. Küpper (New York: Springer Publishing, 2003).

43. See Valencius, *The Lost History of the New Madrid Earthquakes*; and Coen, *The Earthquake Observers*.

44. See Peter U. Rodda and Alan E. Leviton, "Nineteenth Century Earthquake Investigations in California," *Earth Sciences History* 2, no. 1 (1983): 48–56.

45. While Finn focuses on how information circulates after a disaster, her insight that the flow of information related to a crisis can have consequences for how the event, and future events, are experienced and understood is also important here. See Megan Finn, *Documenting Aftermath: Information Infrastructures in the Wake of Disasters* (Cambridge, MA: MIT Press, 2018).

46. Pursuit of knowledge of the underground has not only been important for geoscientists' ongoing disciplinary work but has become key to the territorialization of sovereignty. See Bruce Braun, "Producing Vertical Territory: Geology and Governmentality in Late Victorian Canada," *Ecumene* 7, no. 1 (2000): 7–46; Anthony Bebbington, "Underground Political Ecologies: The Second Annual Lecture of the Cultural and Political Ecology Specialty Group of the Association of American Geographers," *Geoforum* 43, no. 6 (2012): 1152–1162. Stuart Elden, "Secure the Volume: Vertical Geopolitics and the Depth of Power," *Political Geography* 34 (2013): 35–51.

47. On mechanical objectivity, see Lorraine Daston and Peter Galison, *Objectivity* (New York: Zone Books, 2007). On the ways that human and mechanized senses were both used in earthquake research in the nineteenth and twentieth centuries, see Coen, *The Earthquake Observers*.

48. See Paul Edwards, *A Vast Machine: Computer Models, Climate Data, and the Politics of Global Warming* (Cambridge, MA: MIT Press, 2010); and "Entangled Histories: Climate Science and Nuclear Weapons Research," *Bulletin of the Atomic Scientists* 68, no. 4 (2012): 28–40.

49. See Tahani Nadim, "Blind Regards: Troubling Data and Their Sentinels," *Big Data & Society* 3, no. 2 (2016): 1–6; Jennifer Gabrys, *Program Earth: Environmental Sensing Technology and the Making of a Computational Planet* (Minneapolis: University of Minnesota Press, 2016).

50. Key projects related to quickly identifying earthquakes and defining them from background noise are documented in W. H. K. Lee, R. E. Bennett, and K. L. Meagher, *A Method of Estimating Magnitude of Local Earthquakes from Signal Duration* (US Department of the Interior, Geological Survey, 1972); Rex V. Allen, "Automatic Earthquake Recognition and Timing from Single Traces," *Bulletin of the Seismological Society of America* 68, no. 5 (1978): 1521–1532. T. V. McEvilly and E. L. Majer, "ASP: An Automated Seismic Processor for Microearthquake Networks," *Bulletin of the Seismological Society of America* 72, no. 1 (1982): 303–325.

51. The development of the first techniques to estimate a quake's eventual magnitude with data collected at the moment it commences are recorded in Yutaka Nakamura, "On the Urgent Earthquake Detection and Alarm System (UrEDAS)," in *Proceedings of the 9th World Conference on Earthquake Engineering VII* (1988), 673–678; Juan Manuel Espinosa-Aranda, Arturo Uribe, Gerardo Ibarrola, Victor Toledo, and Cecelio Rebollar, "Evaluación de un algoritmo para detectar sismos de subducción," In *Memorias VIII Congreso Nacional de Ingeniería Sísmica* (1989): A199–A211; and Juan Manuel Espinosa-Aranda, Alejandro Jimenez, O. Contreras, Gerardo Ibarrola, and R. Ortega, "Mexico City Seismic Alert System," in *Simposio Internacional Sobre Prevencion de Desastres Sismicos* (Mexico City: CENAPRED, 1992).

52. No European system is public.

53. See, for example, Sarah E. Minson, Annemarie S. Baltay, Elizabeth S. Cochran, Thomas C. Hanks, Morgan T. Page, Sara K. McBride, Kevin R. Milner, and Men Andrin Meier, "The Limits of Earthquake Early Warning Accuracy and Best Alerting Strategy," *Scientific Reports* 9, no. 1 (2019): 1–13; and David Wald, "Practical Limitations of Earthquake Early Warning," *Earthquake Spectra* 36, no. 3 (2020): 1412–1447.

54. Comisión Nacional de Reconstrucción, *Bases para el establecimiento de una sistema nacional de protección civil* (Mexico City: CNR, 1986), 16. See also Organisation for Economic Co-operation and Development. "Mexico 2013: Review of the Mexican National Civil Protection System." *OECD Reviews of Risk Management Policies* (Paris: OECD, 2013). Civil Protection has been institutionalized differently across Mexico. There are Ayuntamientos, Coordinationes, Direcciones, Unidades, Institutos, Subsecretarias, and Secretarias in various states' governments. Each of these implies different levels of access to leaders and of influence in decision-making. They are listed in roughly ascending order here. Funding and personnel also vary, though not necessarily in accordance with the status of Civil Protection offices or with respect to with the risk related to the hazards that the populations of these states might live with.

55. A security apparatus is a matter of epistemological, material, and administrative elements and agents working in coordination. Large-scale coordinated systems can make possible vast integrated configurations of power and sometimes violence in the intimate life of ordinary people. As Stephen Collier points out, Foucault gives us the tools to think through disaster prevention in terms of a "system of correlations"

and "heterogenous ensembles." See Stephen J. Collier, "Enacting Catastrophe: Preparedness, Insurance, Budgetary Rationalization," *Economy and Society* 37, no. 2 (2008): 224–250. See also Masco, *The Theater of Operations*; Tracy C. Davis. *Stages of Emergency: Cold War Nuclear Civil Defense* (Durham, NC: Duke University Press, 2007); Ben Anderson, "Preemption, Precaution, Preparedness: Anticipatory Action and Future Geographies," *Progress in Human Geography* 34, no. 6 (2010): 777–798; Didier Fassin and Mariella Pandolfi, eds., *Contemporary States of Emergency: The Politics of Military and Humanitarian Interventions* (New York: Zone Books, 2010); and Brian Massumi, "Fear (The Spectrum Said)," *Positions* 13, no. 1 (2005): 31–48.

56. The Mexico City authority is, in the schema of Mexico's three-level system of government, a city with the power of a state (and its leader is, consequently, referred to as a "governor") while its various boroughs are classed as municipal governments with the attendant powers and regulatory responsibilities. There are, then, thirty-two entities treated like "states" in Mexico but only thirty-one states.

57. Each of these implies different levels of access to leaders and of influence in decision-making; they are listed in roughly ascending order here. Funding and personnel also vary, though not necessarily in accordance with the status of Civil Protection offices or with respect to with the risk of "perturbatory phenomena" or disastrous fallout from encounters with such hazards that the populations of these states might live with.

58. Guerrero's annual state budget: $5 billion pesos in 2014 when I visited, or approximately $400 million USD. Oaxaca's was $8 billion pesos or $600 million USD. A rich state like Nuevo León might have a budget of around twice that of these two.

59. Quote from interview. Translation by author.

60. Michael Fischer, "Culture and Cultural Analysis as Experimental Systems," *Cultural Anthropology* 22, no. 1 (2007): 3.

61. Stefan Helmreich has discussed this issue thoughtfully in "After Culture: Reflections on the Apparition of Anthropology in Artificial Life, a Science of Simulation," *Cultural Anthropology* 16, no. 4 (2001): 612–627. After spending time in Mexico, I developed a sense that even outside of the social sciences I could expect the term to carry different kinds of meanings there: to describe how people lived and how they made sense of the world or perhaps their traditions, stories, and craftwork. The "cultures" that have been the topic of academic and popular interest are often indigenous; for a critical account of this movement in national mythmaking, see Claudio Lomnitz Adler, *Exits from the Labyrinth: Culture and Ideology in the Mexican National Space* (Berkeley: University of California Press, 1992); and Roger Bartra, *The Cage of Melancholy: Identity and Metamorphosis in the Mexican Character* (New Brunswick, NJ: Rutgers University Press, 1992).

62. This is evident in a number of official statements from this time period, including a public talk by the general coordinator of Mexico City's Program for the Prevention

of Risks and Civil Protection. See José Antonio Carranza Palacios, "La preventción y la cultura sísmica," First National Conference on Civil Protection, December 8–10, (Mexico City: Sistema Nacional de Protección Civil, 1993), 92. See also Elia Arjonilla and Virginia García Acosta, "Cultura sísmica," in *Seminario aprovechamiento del sistema de alerta sísmica* (Mexico City: Fundacion Javier Barros Sierra, 1992).

63. Comisión Nacional de Reconstrucción, *Bases para el establecimiento de una sistema nacional de protección civil*, 101.

64. Ley de Protección Civil 2012.

65. Some suggested that the Japanese had such a culture, when I asked. In the wake of the disaster related to Hurricane Katrina, at least one Civil Protection official was skeptical of the culture of prevention in the United States.

66. See Comisión Nacional de Reconstrucción, *Bases para el establecimiento*.

67. On the effects that such models can have for policy, see Sheila Jasanoff, "A Mirror for Science," *Public Understanding of Science* 23, no. 1 (2014): 21–26.

68. There are many resonances here with recrimination about the so-called "culture of poverty." This model associates certain ways of being with failures to thrive in the world and tends to ignore the forms of dominance that produce structural inequality and affect material conditions of life. See Michele Lamont, Mario Luis Small, and David J. Harding, "Reconsidering Culture and Poverty," *The Annals of the American Academy of Political and Social Science* 629, no. 1 (2010): 6–27. I thank Elizabeth F. S. Roberts for pointing out this relation to me in conversation, and for her work on the topic.

69. Julie Koppel Maldonado, "Considering Culture in Disaster Practice," *Annals of Anthropological Practice* 40, no. 1 (2016): 52–60 uses the term "scapegoating," as do Faas and Barrios, "Applied Anthropology of Risks, Hazards, and Disasters."

70. Jesus Manuel Macías Medrano, *Desastres y protección civil: Problemas sociales, políticos y organizacionales* (Mexico City: CIESAS, 1999), 7. Doing so would articulate a kind of personalization of responsibility that resonates with the security rationalities documented in Davis, *Stages of Emergency*, and Masco, *The Theater of Operations*, with respect to the Cold War and the War on Terror, respectively.

71. Quote from interview. Translation by author.

72. Alex Padilla, "Opening Remarks on Bringing EEW to the US," *Third International Conference on Earthquake Early Warning*, September 3, 2014.

CHAPTER 4

1. Mirna Servín Vega, "Falsa alarma de sismo provoca estampida en inmuebles de la capital," *La Journada*, July 28, 2014, https://www.jornada.com.mx/2014/07/29 /capital/030n1cap.

2. This has since changed as other technoscientists have developed networks of small sensors and even smartphones to use for this purpose. Additionally, note that

I am not referring to on-site devices that can register P-waves of earthquakes and produce alerts before slower and stronger S-waves arrive. While some refer to those as early warning systems too, the CIRES team just calls them "pendulums." They do not rely on complex environmental monitoring networks in the same way that the systems under discussion here do.

3. This is, of course, not always the case. The tragic earthquake of September 19, 2017, originated very near to Mexico City, so the system was only able to give city residents between five and ten seconds of warning. In cases where an earthquake originates on the western coast of Mexico, however, a minute is a conservative estimate of the lead-time that an earthquake early warning system can offer.

4. It was ASCII, or American Standard Code for Information Interchange.

5. Astute readers will note that "crying wolf" was a concern with the alert described in chapter 2 as well. See Elizabeth Reddy, "Crying 'Crying Wolf' How Misfires and Mexican Engineering Expertise are Made Meaningful," *Ethnos* 85, no. 2 (2020): 335–350. J. L. Austin considered misfires as only one ordinary kind of "infelicity" that might interfere with the performative efficacy of a speech act. *How to Do Things with Words* (Oxford: Oxford University Press, 1961).

6. For a more extended and nuanced consideration of engineers and professional dynamics around responsibility, see Jessica M. Smith, *Extracting Accountability: Engineers and Corporate Social Responsibility* (Cambridge, MA: MIT Press, 2021).

7. Dennis S. Mileti and John H. Sorensen, *Communication of Emergency Public Warnings: A Social Science Perspective and State-of-the-Art Assessment* (Washington, DC: Federal Emergency Management Agency, 1990). See also Jeannette Sutton and Erica D. Kuligowski, "Alerts and Warnings on Short Messaging Channels: Guidance from an Expert Panel Process," *Natural Hazards Review* 20, no. 2 (2019): 1–10; and Michele M. Wood, Dennis S. Mileti, Hamilton Bean, Brooke F. Liu, Jeannette Sutton, and Stephanie Madden, "Milling and Public Warnings," *Environment and Behavior* 50, no. 5 (2018): 535–566.

8. See Jeanette Sutton, Sarah C. Vos, Michele M. Wood, and Monique Turner, "Designing Effective Tsunami Messages: Examining the Role of Short Messages and Fear in Warning Response," *Weather, Climate, and Society* 10, no. 1 (2018): 75–87; Jeannette Sutton, Laura Fischer, Lori E. James, and Sarah E. Sheff, "Earthquake Early Warning Message Testing: Visual Attention, Behavioral Responses, and Message Perceptions," *International Journal of Disaster Risk Reduction* 49 (2020); S. K. McBride, A. Bostrom, J. Sutton, R. M. de Groot, A. S. Baltay, B. Terbush, P. Bodin, M. Dixon, E. Holland, R. Arba, P. Laustsen, S. Liu, and M. Vinci, "Developing Post-Alert Messaging for ShakeAlert, the Earthquake Early Warning System for the West Coast of the United States of America," *International Journal of Disaster Risk Reduction* 50 (2020): 101713.

9. See Alissa Walker, "Earthquake-Prone LA Needs a Better Early Warning Alert," *Curbed Los Angeles*, July 10, 2019, https://la.curbed.com/2019/7/10/20688797/earthquake-early-warning-los-angeles-app.

10. See, for example, "Chinese Researchers Build World-Leading Earthquake Early Warning System," *Sputnik*, June 27, 2019, https://sputniknews.com/analysis /201906271076062345-chinese-researchers-world-leading-earthquake-early-warning -system.

11. Mexico had roughly 33.3 million smartphone users at this time according to the Groupe Speciale Mobile Association. *Mobile Economy: Latin America* (London: GSMA, 2013) http://www.gsma.com/latinamerica.

12. Sixty-three percent of Mexicans had access to smartphones in 2018. Groupe Speciale Mobile Association, *The Mobile Economy: Latin America 2019* (London: GSMA, 2019) http://www.gsma.com/latinamerica.

13. As of this writing, these are options for subscribers who pay $7.49 USD per year.

14. Quote from interview. Translation by author.

15. On Mexico's digital frontier, see Héctor Beltrán, "Code Work: Thinking with the System in Mexico," *American Anthropologist* 122, no. 3 (2020): 487–500. This description of the SkyAlert app is very much in keeping with the scholar Lilly Irani's insights about how entrepreneurial ventures are often described. Irani shows that entrepreneurialism is often vaguely defined and understood to be pluripotent. It is expected to "construct markets, produce value, and do nation building" all at once. Lilly Irani, *Chasing Innovation: Making Entrepreneurial Citizens in Modern India* (Princeton, NJ: Princeton University Press, 2019), 2.

16. Quote from interview. Translation by author.

17. Those who are not served by an infrastructure never rely on its function, either. See Susan Leigh Star, "The Ethnography of Infrastructure," *American Behavioral Scientist* no. 3 (1999): 377–391.

18. Infrastructural changes and failures are, of course, never out of the ordinary (see, for example, Nikhil Anand's excellent infrastructural scholarship in *Hydraulic City: Water and the Infrastructures of Citizenship in Mumbai* (Durham, NC: Duke University Press, 2017), or the overview in Casper Bruun Jensen and Atsuro Morita, "Introduction: Infrastructures as Ontological Experiments," *Ethnos* 82, no. 4 (2017): 615–626. Systems break down all the time, and breakdowns must be part of any analysis related to them.

19. Brian Larkin, "The Politics and Poetics of Infrastructure," *Annual Review of Anthropology* 42, no. 1 (2013): 329.

20. Mirna Servín Vega, "Falsa alarma de sismo provoca estampida en inmuebles de la capital."

21. One tweet summarized various interactions after Mexico's July 2014 misfire in rude text transposed over a popular image of a bellicose anteater with its claws spread wide and aggressive: "Ven puto/vente a pelear cabron" it read: "Come at me, pussy/ you bastard, come and fight." It was not clear to me which of the organizations

was meant to be the violent mammal with its claws stretched wide and which the taunted audience. It seemed significant, though, that this gloss of the organizations' interaction is brimming with energy and anticipating action, responding to one event and preparing for a new one.

22. Quote from Twitter. Translated by author.

23. Tom Boellstorff, *Coming of Age in Second Life: An Anthropologist Explores the Virtually Human* (Princeton, NJ: Princeton University Press, 2008).

24. See Denise Maerker, "App SkyAlert 'replicó' falsa alerta sísmica," *Grupo Formula*, July 28, 2014, https://www.radioformula.com.mx/noticias/20140728/app-skyalert-replico-falsa-alerta-sismica-con-denise-maerker/#sthash.s17KNgpW.dpuf.

25. A smartphone app was limited by Mexico's telephone infrastructure, which didn't facilitate broadcast texts but might still warn more people of an impending earthquake than a radio or television message that they had no way of receiving. This is a "Late Alert" in the typology developed by S. K. McBride, A. Bostrom, J. Sutton, R. M. de Groot, A. S. Baltay, B. Terbush, P. Bodin, P. M. Dixon, E. Holland, R. Arba, P. Laustsen, S. Liu, and M. Vinci, "Developing Post-Alert Messaging for ShakeAlert, the Earthquake Early Warning System for the West Coast of the United States of America," *International Journal of Disaster Risk Reduction* 50 (2020).

26. These are sold at ever-more-affordable but still prohibitive costs (as of June 2015, the price might range from $1,300 for a new radio and installation to $100 USD for a used off-brand receiver, though a home receiver to be priced at about $25 is currently in the works).

27. As Stephanie C. Kane, Eden Medina, and Daniel M. Michler note in the course of their reflections on communication failures in Chile following the 2010 Maule earthquake, "Disaster preparation has not guided the development of digital communication infrastructure to the extent required for effective disaster response." Stephanie C. Kane, Eden Medina, and Daniel M. Michler, "Infrastructural Drift in Seismic Cities," *Social Text* 33, no. 1 (2015): 71

28. Narratives that relate technical efforts to the common good of Mexico and its people have powerful implications for how experts consider their social roles and the subjectivities they embody today. Mexican technical experts have a unique position of obligation and ability for public safety. Many feminist science and technology studies scholars have explored the relationship between having the capacity to take some action and feeling an obligation to do so using the term "response-ability." The concept of response-ability directs attention to how ways of knowing the self and the world can constitute the conditions of possibility for action, and vice versa. As Donna Haraway puts it, response-ability is the "praxis of care and response." See Donna Haraway, "Awash in Urine: DES and Premarin® in Multispecies Response-Ability," *WSQ: Women's Studies Quarterly* 40, no. 1 (2012): 301–316. See also Eva Hayward, "Fingeryeyes: Impressions of Cup Corals," *Cultural Anthropology* 25, no. 4 (2010): 577–599;

Astrid Schrader, "Responding to Pfiesteria Piscicida (the Fish Killer): Phantomatic Ontologies, Indeterminacy, and Responsibility in Toxic Microbiology," *Social Studies of Science* 40, no. 2 (2010): 275–306; Aryn Martin, Natasha Myers, and Ana Viseu, "The Politics of Care in Technoscience," *Social Studies of Science* 45, no. 5 (2015): 625–641.

29. Quotes from interview. Translation by author.

30. Mirna Servín Vega, "Falsa alarma de sismo provoca estampida en inmuebles de la capital."

31. El Economista, "SkyAlert acepta error tras alerta sísmica," *El Economista*, July 28, 2014, https://www.eleconomista.com.mx/politica/SkyAlert-acepta-error-tras-alerta -sismica-20140728-0012.html. The news story was picked up by international media too—see *BBC News*, "Mexico quake app firm SkyAlert sorry for false alarm," July 29, 2014, http://www.bbc.com/news/world-latin-america-28543663.

32. Quote from interview. Translation by author.

33. This should not be understood to be a paradox. As Michelle Murphy wrote, "the exercise of power operates through care." Michelle Murphy, "Unsettling Care: Troubling Transnational Itineraries of Care in Feminist Health Practices," *Social Studies of Science* 45, no. 5 (2015): 719; see also Maria Puig de la Bellacasa, *Matters of Care: Speculative Ethics in More Than Human Worlds* (Minneapolis: Minnesota University Press, 2017). Care is not simple and must be interrogated with respect to its material and symbolic attributes and effects.

34. Claudio Lomnitz, "Ritual, Rumor and Corruption in the Constitution of Polity in Modern Mexico," *Journal of Latin American Anthropology* 1, no. 1 (1995): 38. The social work of conspiracy theories is a topic of some interest to anthropologists outside of Mexican contexts. While Lomnitz sees chisme as an important form of national discourse and crucial to the national public, Charles Briggs has documented the conspiracy theories that such gossip may contain as political epistemological practices of marginalized communities. Charles Briggs, "Theorizing Modernity Conspiratorially: Science, Scale, and the Political Economy of Public Discourse in Explanations of a Cholera Epidemic," *American Ethnologist* 31, no. 2 (2004): 164–187. Hoon Song points out how such theories, and our theorizing about them, might trouble our understandings of knowledge. Hoon Song, "Cogito, Mimesis, and Conspiracy Theory," *Culture, Theory and Critique* 53, no. 1 (2012): 1–18.

35. See John Gledhill, *Power and its Disguises: Anthropological Perspectives on Politics* (Boulder, CO: Pluto Press, 1994); Pieter De Vries, "Vanishing Mediators: Enjoyment as a Political Factor in Western Mexico," *American Ethnologist* 29, no. 4 (2002): 901–927.

36. In her recent ethnography of Guatemalan conservation work, Micha Rahder describes similar theory work as a "paranoid epistemology," a "mode of thought borne of social contexts in which suspicion of hidden dealings is a perfectly

reasonable response." *An Ecology of Knowledges: Fear, Love and Technoscience in Guatemalan Forest Conservation* (Durham, NC: Duke University Press, 2020): 27.

37. Lomnitz, "Ritual, Rumor and Corruption in the Constitution of Polity in Modern Mexico," 36.

38. SkyAlert 4, *System Information* (Accessed July 1, 2020).

CHAPTER 5

1. Quote from interview. Translation by author.

2. Quote from interview. Translation by author.

3. See Ethan Blue, Michal Levine, and Dean Nieusma, *Engineering and War: Militarism, Ethics, Institutions, Alternatives* (Williston, VT: Morgan and Claypool, 2013) for a discussion of engineering as a domain of knowledge, a set of practices, a profession, and an ideology.

4. See Wendy Faulkner, "'Nuts and Bolts and People': Gender-Troubled Engineering Identities," *Social Studies of Science* 37, no. 3 (2007): 331–356.

5. Quote from interview. Translation by author.

6. For an extended and thoughtful historical meditation on interplay of materials, engineering knowledge, and broad social trends, see Amy E. Slaton, *Reinforced Concrete and the Modernization of American Building, 1900–1930* (Baltimore, MD: Johns Hopkins University Press, 2001).

7. On measurement practices and their social worlds as objects of anthropological inquiry, see also Antonia Walford, "Double Standards: Examples and Exceptions in Scientific Metrological Practices in Brazil," *Journal of the Royal Anthropological Institute* 21, no. S1 (2015): 64–77.

8. Atsushi Akera, *Calculating a Natural World* (Cambridge, MA: MIT Press, 2007).

9. Although one should not assume that one epistemology alone informs the technical development of a system simply because those directing design and construction were trained in it. See Chandra Mukerji, *Impossible Engineering: Technology and Territoriality on the Canal Du Midi* (Princeton, NJ: Princeton University Press, 2012).

10. In a survey of thirty-two of the seventy-four CIRES employees at the time of my fieldwork, only six identified themselves as engineers—that is, people who had completed the education necessary to attain the title of engineer and were employed in that role. Others were still contributing to engineering projects, even without the title and position.

11. On the topic of engineering identities, see Karen L. Tonso, *On the Outskirts of Engineering: Learning Identity, Gender, and Power via Engineering Practice* (Rotterdam: Sense Publishers, 2007). She and other scholars of engineering have emphasized

questions about who can be an engineer that I do not go into here, but interested readers should also see Amy E. Slaton, *Race, Rigor, and Selectivity in U.S. Engineering: The History of an Occupational Color Line* (Cambridge, MA: Harvard University Press, 2010). Some researchers now describe knowledge traditions of marginalized and non-Western communities as engineering to better respect the expert technical practice that they involve. See, for example, the decolonial work of Joel A. Mejia and Alberto López Pulido, "Fregados Pero no Jodidos: A Case Study of Latinx Rasquachismo," in *American Society for Engineering Education Annual Conference Proceedings* (2018); and Gordon D. Hoople, Joel A. Mejia, Diana A. Chen, and Susan M. Lord, "Reimagining Energy: Deconstructing Traditional Engineering Silos Using Culturally Sustaining Pedagogies," in *American Society for Engineering Education Annual Conference Proceedings* (2018).

12. See Gary Lee Downey and Juan C. Lucena, "National Identities in Multinational Worlds: Engineers and 'Engineering Cultures,'" *International Journal of Continuing Engineering Education and Life-Long Learning* 15 (2005).

13. Blue, Levine, and Nieusma, *Engineering and War*; Juan C. Lucena, *Defending the Nation: US Policymaking to Create Scientists and Engineers from Sputnik to the "War Against Terrorism"* (Lanham: University Press of America, 2005); David Noble, *America by Design: Science, Technology, and the Rise of Corporate Capitalism* (Oxford: Oxford University Press, 1979) details a history of engineering, capitalism, and the armed forces. See Scott Knowles, *The Disaster Experts: Mastering Risk in Modern America* (Philadelphia: University of Pennsylvania Press, 2012) for a consideration of how urbanization and different forms of life as well as capital have driven movements of professionalization and standardization of engineers in the United States. These projects, of course, have been far from the only visions structuring the field: see Matthew Wisnioski, *Engineers for Change: Competing Visions of Technology in 1960s America* (Cambridge, MA: MIT Press, 2012).

14. Luz Fernanda Azuela and Rafael Guevara Fefer offer an account of how writing about of these movements in science and technology has been strongly motivated by modernist, developmentalist narratives. See L. Azuela and R. Guevera Fefer, "La ciencia en México en el siglo XIX: Una aproximación historográfica," *Asclepio* 2 (1998).

15. These "arts" meant "scientific and technical instruction for mine owners, metallurgists, engravers, draftsmen, engineers, architects, farmers, druggists, seamen, artists, and other artisans," and schools often drew on financial support of interested parties, including mine owners and merchants who might employ graduates. Juan José Saldaña, *Science in Latin America* (Austin: University of Texas Press, 2006), 53. See also J. J. Izquierdo, "La primera casa de la ciencias en México: El Real Seminario de Minería 1792–1811" (México: Ediciones Ciencia, 1958); Luz Fernanda Azuela, "La geología en la formación de los ingenieros Mexicanos del siglo XIX," in *Formación de ingenieros en el México del siglo XIX*, ed. M. Ramos Lara and R. Rodríguez Benítez

(Mexico City: Universidad Nacional Autónoma de México Press, 2007), 91–108; Dogan and Parhe, *Las nuevas ciencias sociales: La marginalidad creadora*, (México: Grijalbo, 1991); and P. Arias, *Industria y estado en la vida de México* (Zamora: El Colegio de Michoacán, 1990). This focus on useful arts should be understood in the frame of Mexico's position on the "periphery" of European imperial power and science.

16. On the history of Mexican engineering and science education during its time as a Spanish colony, much has been written. Scholars concerned with the sweeping history of Mexican science and technology interests offer good accounts, including E. Gortari, *La ciencia en la historia de México* (Mexico City: FCE, 1963) J. J. Saldaña, *Science in Latin America* (Austin: University of Texas Press, 2006). For detailed investigations of key moments and communities, the following may be of use: W. Howe, *The Mining Guild of New Spain and Its Tribunal General, 1770–1821* (Greenwood Press, 1968); K. Brown, *A History of Mining in Latin America from the Colonial Era to the Present* (Albuquerque: University of New Mexico Press, 2012); D. A. Brading, "Miners and Merchants in Bourbon Mexico 1763–1810" (Cambridge: Cambridge University Press, 1971).

17. On the history of Mexican engineering in the nineteenth century, see M. Ramos Lara and R. Rodríguez Benítez, eds., *Formación de ingenieros en el México del siglo XIX* (Mexico City: Universidad Nacional Autónoma de México Press, 2007); M. Ramos Lara, *Vicisitudes de la ingeniería en México (Siglo XIX)* (Mexico City: Universidad Nacional Autónoma de México Press, 2013); M. Bazant, *Historia de la educacion durante El Porfiriato* (Mexico City: El Colegio de Mexico, 2014). As Juan Lucena points out, Mexican engineering has been institutionalized around ideas of public well-being that are particularly related to state projects. See Juan C. Lucena, "De Criollos a Mexicanos: Engineers' Identity and the Construction of Mexico," *History and Technology* 23, no. 3 (2007): 275–288.

18. Jacqueline Fortes and Larissa Adler Lomnitz, *Becoming a Scientist in Mexico: The Challenge of Creating a Scientific Community in an Underdeveloped Country* (University Park: University of Pennsylvania Press, 1994).

19. The son of a composer and brother of a high-ranking career Institutional Revolutionary Party (commonly called the PRI) functionary who served as minister of foreign affairs and minister of finance, Carrillo Flores (1911–1967) was a mid-century soil mechanics and atomic energy researcher. He earned degrees from UNAM and then MIT in the United States before serving as Mexico's representative at Bikini Atoll and, eventually, rector at UNAM. Historians Gisela Mateos and Edna Suarez-Díaz argue that his international circulations were crucial for Mexico's emergent nuclear research program. For more on Carrillo Flores's international circulations, see Gisela Mateos and Edna Suárez-Díaz, "Peaceful Atoms in Mexico," *Beyond Imported Magic* (Cambridge, MA: MIT Press, 2014).

20. Rosenblueth (1926–1994) was a civil engineer who specialized in seismic safety in the mid- and late-twentieth century. Trained at UNAM and then at the University

of Illinois, at various times he served as a subminister of public education, consulted for UNESCO, and played a leadership role in many engineering organizations. His work was not limited to engineering, though; his published work refers to continental philosophy and art as well as Mexican intellectual traditions, and explicitly engages with issues related to race and culture. His father was a well-recognized artist, and his cousin, a medical doctor, coauthored work on cybernetics with Norbert Weiner and participated in the Macy Conferences.

21. Still, only 51 percent of Mexican adults have graduated from secondary school, and only 22 percent have any form of higher degree at all. These figures refer to adults between the ages of twenty-five and sixty-four, as reported in the Secretaria de Educación Pública, *Principales cifras del sistema educativo nacional 2015–2016* (Mexico City: Dirección General de Planeación, Programación y Estadística Educativa, 2016).

22. More than 700,000 students enrolled in engineering programs in 2011, almost double the number who enrolled in 2000. There was significant growth in education related to technology during this time, but in each of these years, engineering students comprised over 70 percent of the total population of students matriculating. See Academia de Ingeniería, *Menú educación-general. Recuperado el 5 de 2014, de Observatorio de la Ingeniería Mexicana* (Mexico City: Academia de Ingeniería, 2012); and Academia de Ingeniería, *Coloquio Sobre Formación de Ingenieros en México* (Mexico City: Academia de Ingeniería, 2016), 8.

23. Academia de Ingeniería, *Coloquio sobre formación de ingenieros en México*.

24. Roderic A. Camp, *Politics in Mexico: The Decline of Authoritarianism* (Oxford: Oxford University Press, 1999).

25. While "technoscience" is a fairly common term in the fields of STS and cultural anthropology, I use it here in light of the work of two key thinkers. Bruno Latour suggests that it allows us to critically consider what is understood to be "inside" and "outside" of the scope of science and engineering and query boundaries between science and society. See Bruno Latour, *Science in Action: How to Follow Scientists and Engineers through Society* (Cambridge, MA: Harvard University Press, 1987), 174 and 176. Donna Haraway builds on this work to consider the messy and unexpected ways that human and nonhuman actors are brought into relation around the topic. She is particularly interested in how power is enacted in these contexts. See Donna Haraway, *ModestWitness@SecondMillennium.FemaleManMeetsOncoMous: Feminism and Technoscience* (New York: Routledge, 1997), 50–51. While technoscience is, in their hands, somewhat all-encompassing, other thinkers argue that it is still useful to attend to technoscience's alternatives and margins. For example, Arun Agrawal has written powerfully about the ways that indigenous environmental knowledge is contrasted to scientific or Western environmental knowledge in scholarship and development practice. See Arun Agrawal, "Dismantling the Divide Between Indigenous and Scientific Knowledge," *Development and Change* 26, no. 3 (1995): 413–439. Distinguishing these forms of knowledge from each other can, as Povinelli points

out, be crucial to distributing power and authority in late liberalism. See Elizabeth Povinelli, *Geontologies: A Requiem to Late Liberalism* (Durham, NC: Duke University Press, 2016).

26. Historian Amy Slaton notes that this narrative can be misleading and historically inaccurate but is nonetheless indicative of powerful narratives in the field Amy E. Slaton, *Reinforced Concrete and the Modernization of American Building, 1900–1930* (Baltimore, MD: The Johns Hopkins University Press, 2001), 21.

27. Fortes and Lomnitz, *Becoming a Scientist in Mexico*, 2.

28. On engineering jokes and how they can be understood to highlight values and practices related to disciplinary identities, see Donna Riley, *Engineering and Social Justice* (Williston, VT: Morgan and Claypool, 2008).

29. Downey writes about this as a matter of *agencies of construction* and *response*. This language particularly comes from studies of US engineering students who were engaged in identity work around their chosen field, but I find it appropriate for this application too. See Gary Lee Downey, "What Is Engineering Studies for? Dominant Practices and Scalable Scholarship," *Engineering Studies* 1, no. 1 (2009): 55–76.

30. Helen Verran, "Number as an Inventive Frontier in Knowing and Working Australia's Water Resources," *Anthropological Theory* 10, no. 1–2 (2010): 171.

31. As Star and Lampland put it, attention to "boring things" like measurement can allow us insight into important social relations and structures that inform and are shaped by "their historical development, their political consequences, and the . . . decisions made about them." Martha Lampland and Susan Leigh Star, *Standards and Their Stories: How Quantifying, Classifying, and Formalizing Practices Shape Everyday Life* (Ithaca, NY: Cornell University Press, 2009), 13. See also Theodore Porter, *Trust in Numbers: The Pursuit of Objectivity in Science and Public Life* (Princeton, NJ: Princeton University Press, 1995); Jane I. Guyer, Naveeda Khan, and Juan Obarrio, "Number as an Inventive Frontier," *Anthropological Theory* 10, no. 1 (2010): 63–86.

32. Diana Forsythe found a similar description of epistemic work in a computer science lab in the late 1990s. There, Forsythe observed an engineering ethos marked by what she calls "a tendency to approach things practically rather than theoretically." She quotes an informant: "if you waited to figure out what you were doing, you'd never get anything done. There just isn't time to do that." Diana Forsythe, *Studying Those Who Study Us: An Anthropologist in the World of Artificial Intelligence* (Stanford, CA: Stanford University Press, 2001), 44.

33. For curious readers, the particular operations that they performed involved taking derivatives of readings and performing comparative topographical analyses. Duran noted a resonance between this analysis and super vector theory, books about which were packed onto the shelves in his overcrowded office, but also explained that this was coincidental and that this form of mathematical analytic was not available to Espinosa Aranda in the 1980s.

34. J. M. Espinosa-Aranda, A. Jiménez, O. Contreras, G. Ibarrola, and R. Ortega, "Mexico City Seismic Alert System," *International Symposium on Earthquake Disaster Prevention* (Mexico City: CENAPRED, JICA, IDNDR, 2000).

35. This is, of course, an explanation for a general audience, written to draw attention to particular features of data analysis. A more technical one can be found in Juan Manuel Espinosa-Aranda, Armando Cuellar, Armando Garcia, Gerardo Ibarrola, Roberto Islas, Samuel Maldonado, and F. H. Rodriguez, "Evolution of the Mexican Seismic Alert System (SASMEX)," *Seismological Research Letters* 80, no. 5 (2009): 694–709; or A. Cuéllar, J. M. Espinosa-Aranda, G. Suárez, G. Ibarrola, A. Uribe, F. H. Rodríguez, R. Islas, G. M. Rodríguez, A. Garcia, and B. Frontana, "The Mexican Seismic Alert System (SASMEX): Its Alert Signals, Broadcast Results and Performance During the M7.4 Punta Maldonado Earthquake of March 20 2012," in *Early Warning for Geological Disasters*, ed. F. Wenzel and J. Zschau (Berlin, Heidelberg: Springer-Verlag, 2014), 307–331.

36. While other innovations made this possible, CIRES's Espinosa Aranda and a Japanese engineer named Nakamura seem to have arrived at this unique application independently in the late 1980s as solutions to their particular problems, as described in chapter 3.

37. For more on engineering as a pragmatics, see Penelope Harvey and Hannah Knox, *Roads: An Anthropology of Infrastructrue and Expertise* (Ithaca, NY: Cornell University Press, 2015).

38. Showing these off was a frequent occurrence. Many engineers and technicians at CIRES had pet projects that could be used to change how the system worked in some small way. Senior management retained power to decide whether to move forward with a project or nix it.

39. See for example, Dominique Vinck, Eric Blanco, Michel Bovy, Pascal Laureillard, Oliver Lavoisy, Stepahe Mer, Nathalie Ravaille, and Thomas Reverdy, *Everyday Engineering: An Ethnography of Design and Innovation* (Cambridge, MA: MIT Press, 2003); and Louis L. Bucciarelli, *Designing Engineers* (Cambridge, MA: MIT Press, 1994).

40. Fortes and Lomnitz, *Becoming a Scientist in Mexico*.

41. A. Iglesias, S. K. Singh, M. Ordaz, M. A. Santoyo, and J. Pacheco, "The Seismic Alert System for Mexico City: An Evaluation of Its Performance and a Strategy for Its Improvement," *Bulletin of the Seismological Society of America* 97, no. 5 (2007): 1727.

42. About twenty-eight seconds for events occurring at epicentral distances of up to 270 kilometers away, as opposed to around sixty seconds offered by the current Sistema de Alerta Sísmica Mexicano. See Iglesias et al., "The Seismic Alert System for Mexico City," 1727.

43. Iglesias et al., "The Seismic Alert System for Mexico City," 1720.

44. Iglesias et al., "The Seismic Alert System for Mexico City," 1728. These would be categorized as "inaccurate" alerts in a recent typography of alerting. See S. K. McBride, A. Bostrom, J. Sutton, R. M. de Groot, A. S. Baltay, B. Terbush, P. Bodin, P. M. Dixon, E. Holland, R. Arba, P. Laustsen, S. Liu, and M. Vinci, "Developing Post-Alert Messaging for ShakeAlert, the Earthquake Early Warning System for the West Coast of the United States of America," *International Journal of Disaster Risk Reduction* 50 (2020).

45. See Espinosa-Aranda et al., "Evolution of the Mexican Seismic Alert System (SASMEX)."

46. This is the very issue that framed the development of a typology of kinds of alerting trouble by McBride, et al., "Developing Post-Alert Messaging for ShakeAlert."

47. Quotes from a public event.

CHAPTER 6

1. Field stations are named with reference to their location, generally after nearby communities.

2. A deep and thoughtful exploration of the role of technicians in knowledge work can be found in Caitlin Wylie, *Preparing Dinosaurs: The Work behind the Scenes* (Cambridge, MA: MIT Press, 2021).

3. See Nicole Staroslieski, *The Undersea Network* (Durham, NC: Duke University Press, 2015).

4. This account, particularly in its focus on knowledge and ever-present threats of violence that engineers and technicians experience in the field, might well be read alongside Micha Rahder, *An Ecology of Knowledges: Fear, Love and Technoscience in Guatemalan Forest Conservation* (Durham, NC: Duke University Press, 2020).

5. On the social, technical, and material existence of infrastructures, with particular attention to when and for whom they become relevant, see Susan Leigh Star and Karen Ruhleder, "Steps toward an Ecology of Infrastructure: Design and Access for Large Information Spaces," *Information Systems Research* 7, no. 1 (1996); Susan Leigh Star, "The Ethnography of Infrastructure," *American Behavioral Scientist* 43, no. 3 (1999); Geoffrey Bowker and Susan Leigh Star, *Sorting Things Out: Classification and Its Consequences* (Cambridge, MA: MIT Press, 1999); Arild Jansen and Petter Nielsen, "Theorizing Convergence: Co-Evolution of Information Infrastructures," *Scandinavian Journal of Information Systems* 17, no. 1 (2005): 67–100. David Ribes and Thomas A. Finholt, "The Long Now of Technology Infrastructure: Articulating Tensions in Development," *Journal of the Association for Information Systems* 10 (2009): 375–398; Geoffrey Bowker, Karen Baker, Florence Millerand, and David Ribes, "Towards Information Infrastructure Studies: Ways of Knowing in a Networked Environment," in *International Handbook of Internet Research*, ed. J. Hunsinger, Li. Klastrup, and M. Allen (New York: Springer, 2010).

6. As critical data scholars frequently remind us, data are always the result of and the occasion for relationships between humans, technologies, institutions, forces, environments, animals, plants, and so on that are absolutely necessary but will never be formally addressed in most discussions of earthquake early warning. See Antonia Walford, "Raw Data: Making Relations Matter," *Social Analysis* 61 (2017): 65–80.

7. Lucy Suchman, "Centers of Coordination: A Case and Some Themes," *Discourse, Tools and Reasoning—Essays on Situated Cognition* 47 (1997); also see also Daniel Neyland, "The Accomplishment of Spatial Adequacy: Analysing CCTV Accounts of British Town Centres," *Environment and Planning D: Society and Space* 24, no. 4 (2006): 599–613.

8. The simplicity of these maps should not let us forget that there is much more going on in earthquake early warning than can be represented here. Calling attention to both action and infrastructure, environmental data scholar Jennifer Gabrys writes: "Environmental data, monitoring practices and technologies undergo complex processes of negotiation and shaping that do not simply translate a phenomenon monitored into a data point." Jennifer Gabrys, "Practicing, Materialising and Contesting Environmental Data," *Big Data & Society* (2016): 2; see also Jennifer Gabrys, *Program Earth: Environmental Sensing Technology and the Making of a Computational Planet.* (Minneapolis: University of Minnesota Press, 2016); and Starosielski, *The Undersea Network.*

9. While interviews indicated that some of their female colleagues had gone to the field in the past, field visits were not a mixed-gender undertaking during my time at CIRES.

10. As Walford writes of field data gathered in a remote site in the Amazon rainforest, "Without the ongoing relational work . . . data would not exist. Its uniqueness is not due to its isolation, as such, but to the very particular relational configuration from which it emerges." *Raw Data*, 73. This insight does not at first seem out of keeping with how STS scholarship, particularly that related to Actor Network Theory (ANT), understands facts to become established (see, for example, Michel Callon, "Some Elements of a Sociology of Translation: Domestication of the Scallops and the Fishermen of St. Brieuc Bay," in *Power, Action and Belief: A New Sociology of Knowledge?*, ed. J. Law (London: Routledge, 1986); or Bruno Latour, *Reassembling the Social. An Introduction to Actor-Network Theory* (Oxford: Oxford University Press, 2005). Walford is not, however, addressing ANT scholarship's frequent preoccupation with how data's significance is stabilized for technoscientific purposes. Instead, her work shows how more-than-technoscientific relations are always embedded in data production. She suggests that these relations can both make data collection possible and also make the data itself problematic. On ecologies of knowledge and practice, also see Geoffrey Bowker, Stefan Timmermans, Adele E. Clark, and Ellen Balka, eds., *Boundary Objects and Beyond: Working with Leigh Star* (Cambridge, MA: MIT Press, 2015).

11. Quote from interview. Translation by author.

12. Quote from interview. Translation by author.

13. Thirty-two of seventy-four CIRES employees participated in this anonymous survey in 2013; of the twelve members of the communications department that sent teams to the field, eleven indicated that they were born and had lived all their lives in Mexico City.

14. Conversations with Adriana Minor Garcia, a historian of science, have been critical for helping me see how this language works to diminish places outside Mexico City. As Cynthia Hewitt de Alcántara's history of research in rural Mexico demonstrates, the differences that make a difference (as we anthropologists say) between rural and urban Mexican spaces are in flux. She describes the body of research devoted to rural Mexico in the twentieth century as a matter of "cumulative—and in some respects, dialectical . . . half-century-long encounter with the Mexican countryside." Cynthia Hewitt de Alcantara, *Anthropological Perspectives on Rural Mexico* (New York: Routledge and Kegan Paul, 1984), 188. Some observers have considered the differences it marks to be so extreme that it became appropriate to discuss them as two civilizations. See, for example, Guillermo Bonfil Batalla, *México Profundo: Reclaiming a Civilization* (Austin: University of Texas Press, 2004), xv. However, grounding this contrast in specific regional logics, peasant and popular movements, and shifting agrarian policies and modes of production offer more nuanced perspectives on the ways these differences are structured. See Claudio Lomnitz-Adler, "Concepts for the Study of Regional Culture," *American Ethnologist* 18 (1991): 195–214; Peter Guardino, *Peasants, Politics, and the Formation of Mexico's National State: Guerrero, 1800–1857* (Stanford, CA: Stanford University Press, 2002); Gilbert Michael Joseph and Daniel Nugent, *Everyday Forms of State Formation: Revolution and the Negotiation of Rule in Modern Mexico* (Durham, NC: Duke University Press, 1994); and Rodger Bartra, *Agrarian Structure and Political Power in Mexico* (Baltimore, MD: The Johns Hopkins University Press, 1993). Andrew S. Mathews, for example, has shown that these distinctions have significant implications for policy and knowledge projects. See Andrew S. Mathews, "Building the Town in the Country: Official Understandings of Fire, Logging and Biodiversity in Oaxaca, Mexico, 1926–2004," *Social Anthropology* 14 (2006): 335–359.

15. Armando Cuéllar, Gerardo Ibarrola Álvarez, C. Samuel Maldonado, and Juan Manuel Espinosa Aranda, "Sistema de Alerta Sísmica para la ciudad de México," *Revista Digital Universitaria* 11, no. 1 (2010): 1–10; and Juan Manuel Espinosa Aranda, Armando Cuéllar, Armando Garcia, Gerardo Ibarrola, Roberto Islas, Samuel Maldonado, and F. H. Rodriguez, "Evolution of the Mexican Seismic Alert System (SASMEX)," *Seismological Research Letters*, 80 (2009): 694–706.

16. These dangers were of serious concern to the CIRES team whose aim was to keep the stations constantly running and ready for any earthquake. See Alejandro Jiménez, Juan Manuel Espinosa, F. Aicántar, and I. Garcia, "Analisis de confiabilidad del Sistema de Alerta Sismica," in *X Congreso Nacional de Ingeniería Sísmica* (Puerto

Vallarta, Jalisco: Sociedad Mexicana de Ingeniería Sísmica, October 8–11, 1993), 629–634.

17. Quote from interview. Translation by author.

18. Quote from interview. Translation by author.

19. Contrast this to what Dalakoglou and Harvey have called the "paradigmatic material infrastructure of the twenty first century." See Dimitris Dalakoglou and Penny Harvey, "Roads and Anthropology: Ethnographic Perspectives on Space, Time and (Im)Mobility," *Mobilities* 7 (2012): 459.

20. Quote from interview. Translation by author.

21. Quote from interview. Translation by author.

22. Quote from interview. Translation by author.

23. A one-time payment of 18,000 pesos, or nearly $1,400 USD, for the ongoing use of the property was not uncommon (exchange rate calculated for 2014, when this interview occurred).

24. Quote from interview. Translation by author.

25. Quote from interview. Translation by author.

26. His name for them resonates with science-fictional uncertainties about identity, origin, and capacity. See Sarah Franklin, *Dolly Mixtures: The Remaking of Genealogy* (Durham, NC: Duke University Press, 2007). They are copies of CIRES field stations, certainly, but they are more than that. Where they come from cannot be precisely known, at least not safely. What kind of system they might be part of, or what kind of work they might have been developed for, is similarly impossible to get any real information about.

27. I use "organized crime" rather than "narcoviolence" here because of the diversity of dangerous and illegal projects that engineers and technicians might have encountered in rural Mexico in 2014. These are not necessarily directly drug related. For example, self-defense leagues are also extremely dangerous. See Patricio Asfura-Heim, and Ralph H. Espach, "The Rise of Mexico's Self-Defense Forces Vigilante Justice South of the Border," *Foreign Affairs*, 2013. However, I also acknowledge that it is difficult to refer to perpetrators of this violence as "criminals" when evidence suggests that their organizations carry out some of the roles and duties of a traditional government, even as they make heavy use of spectacular intimidation and brutality. See George W. Grayson, "La Familia Drug Cartel: Implications for U.S.-Mexican Security," December 2010, *Strategic Studies Institute*; and Andrew Lantz, "The Performativity of Violence: Abducting Agency in Mexico's Drug War," *Journal of Latin American Cultural Studies* 25 (2016): 253–269. Further, observers of Mexico's drug war inside and outside of academia have been increasingly critical of the state's role in this violence. As Alejandro Zagato points out, the Mexican activist slogan "it was the state!" should direct our attention to how violence is enacted and enabled not just by organizations that have taken on state functions but also by the Mexican

state itself. See Alessandro Zagato, "State and Warfare in Mexico," *Social Analysis* 62 (2018): 55–75.

28. See, for example, Francisco Goldman's excellent reporting for the *New Yorker*. Francisco Goldman, "Crisis in Mexico: The Disappearance of the Forty-Three," *New Yorker*, October 2014; Francisco Goldman, "Mexico's Missing Forty-Three: One Year, Many Lies, and a Theory That Might Make Sense," *New Yorker*, September 30, 2015; Francisco Goldman, "The Missing Forty-Three: The Mexican Government Sabotages Its Own Independent Investigation," *New Yorker*, April 2016.

29. As reported in INEGI 2018. Not all can be associated with organized crime, certainly, but the incredible size of the difference between murder rates in 2014 and 2007 is significant. Further, these numbers do not include those whose remains have not been found and those who may be officially considered "missing." Instituto Nacional de Estadística y Geografía (INEGI) Mortalidad: Conjunto de datos: Defunciones por homicidios, 2018, http://www.inegi.org.mx/lib/olap/consulta/general_ver4/MDXQueryDatos.asp?proy=.

30. Viridiana Rios, "Why Did Mexico Become So Violent? A Self-Reinforcing Violent Equilibrium Caused by Competition and Enforcement," *Trends in Organized Crime* 16 (2013): 138–155.

31. Cf. Paul Kockelman, "How to Take Up Residence in a System," *Journal of Linguistic Anthropology* 20 (2010): 406–402.

32. Quote from interview. Translation by author.

33. Quote from interview. Translation by author.

34. Indeed, this was possible on September 8, 2017, but sadly less so on September 19 when a quake emerged from a fault where CIRES had not positioned field stations.

35. When mapping projects explicitly take on politics, it often happens through the creation of what Young and Gilmore call a "politically potent map." They make the case that this approach assumes a "rational and deliberative" system in which claims simply need to be made in the right ways. See Jason Young and Michael P. Gilmore, "The Spatial Politics of Affect and Emotion in Participatory GIS," *Annals of the Association of American Geographers* 103 (2013): 814. Just as a map is not necessarily an accurate reflection of the world, nor do its political effects need to be conceptualized as direct. See also Joe Gerlach, "Editing Worlds: Participatory Mapping and a Minor Geopolitics," *Transactions of the Institute of British Geographers* 40 (2015): 273–286.

CONCLUSION

1. M. Wood, *State-of-the-Art Knowledge of Protective Actions Appropriate for Earthquake Early Warning*, report prepared for the Cascadia Region Earthquake Workgroup (Denver, CO: Nusura, Inc., 2018); S. E. Minson, M.-A. Meier, A. S. Baltay, T. C. Hanks, and E. S. Cochran, "The Limits of Earthquake Early Warning: Timeliness of Ground Motion Estimates," *Science Advances* 4 (2018); O. Kamigaichi, M. Saito, K.

Doi, T. Matsumori, S. Tsukada, K. Takeda, T. Shimoyama, K. Nakamura, M. Kiyomoto, and Y. Watanabe, "Earthquake Early Warning in Japan: Warning the General Public and Future Prospects," *Seismological Research Letters* 80, no. 5 (2009): 717–726; K. A. Porter and J. L. Jones, "How Many Injuries Can Be Avoided in the HayWired Scenario through Earthquake Early Warning and Drop, Cover, and Hold On?," in S. T. Detweiler and A. M. Wein, eds., *The HayWired Earthquake Scenario—Engineering Implications* (U.S. Geological Survey Scientific Investigations Report, 2018).

2. Y. Horiuchi, "Earthquake Early Warning Hospital Applications," *Journal of Disaster Research* 4, no. 4 (2009): 237–241; E. Yamasaki, "What We Can Learn from Japan's Earthquake Warning System," *Momentum* 1, no. 2 (2012): 1–26; Y. Fujinawa and Y. Noda, "Japan's Earthquake Early Warning System on 11 March 2011: Performance, Shortcomings, and Changes," *Earthquake Spectra* 29, no. S1 (2013): S341–S368.

3. H. S. Kuyuk, R. M. Allen, H. Brown, M. Hellweg, I. Henson, and D. Neuhauser, "Designing a Network-Based Earthquake Early Warning Algorithm for California: ElarmS-2," *Bulletin of the Seismological Society of America* 104 (2014): 162–173; D.-H. Sheen, J.-H. Park, H.-C. Chi, E.-H. Hwang, I.-S. Lim, Y. J. Seong, and J. Pak, "The First Stage of an Earthquake Early Warning System in South Korea," *Seismological Research Letters* 88, no. 6 (2017): 1491–1498.

4. Bhanu Pratap Chamoli, Ashok Kumar, Da-Yi Chen, Ajay Gairola, Ravi S. Jakka, Bhavesh Pandey, Pankaj Kumar, and Govind Rathore, "A Prototype Earthquake Early Warning System for Northern India," *Journal of Earthquake Engineering* 25, no. 12 (2021): 2455–2473.

5. J. Clinton, A. Zollo, A. Mărmureanu C. Zulfikar, and S. Parolai, "State-of-the Art and Future of Earthquake Early Warning in the European Region," *Bulletin of Earthquake Engineering* 14, no. 9 (2016): 2441–2445.

6. C. Satriano, L. Elia, C. Martino, M. Lancieri, A. Zollo, and G. Iannaccone, "PRESTo, the Earthquake Early Warning System for Southern Italy: Concepts, Capabilities and Future Perspectives," *Soil Dynamics and Earthquake Engineering* 31 (2010): 137–153; A. Mărmureanu, C. Ionescu, and C. O. Cioflan, "Advanced Real-Time Acquisition of the Vrancea Earthquake Early Warning System," *Soil Dynamics and Earthquake Engineering* 31 (2010): 163–169.

7. An excellent literature review of these and other pieces was developed by Richard M. Allen and Diego Melgar, "Earthquake Early Warning: Advances, Scientific Challenges, and Societal Needs," *Annual Review of Earth and Planetary Sciences* 47 (2019): 361–388.

8. Quote from interview. Translation by author.

9. See, for example, Charles R. Hale, "Activist Research v. Cultural Critique: Indigenous Land Rights and the Contradictions of Politically Engaged Anthropology," *Cultural Anthropology* 21, no. 1 (2006): 96–120; Louise Lamphere, "The Convergence of Applied, Practicing, and Public Anthropology in the 21st Century," *Human Organization* 63, no. 4 (2004): 431–443; Anthony Oliver-Smith, "Disaster Risk Reduction and Applied Anthropology," *Annals of Anthropological Practice* 40, no. 1 (2016): 73–85.

10. Gary Lee Downey, "What Is Engineering Studies For? Dominant Practices and Scalable Scholarship," *Engineering Studies* 1, no. 1 (2009): 55–76. See also Gary Lee Downey and Teun Zuiderent-Jerak, eds., *Making and Doing: Activating STS through Knowledge Expression and Travel* (Cambridge, MA: MIT Press, 2021). I also thank Jessica Smith, Marie Stettler Kleine, and Emily York for helping me think through the significance of such a project.

11. Annie Y. Patrick, *Engaging with the Invisible: STS Groundwork in an Electrical and Computer Engineering Department* (PhD Diss., Virginia Tech, 2021).

12. Emily Martin, "Meeting Polemics with Irenics in the Science Wars," *Social Text* 46/47 (1996): 57. See also the discussion in Stefan Helmreich, "Nonlinear Works and Lives," presented at the Annual Meeting of the American Anthropological Association, Philadelphia, PA, December 2–6, 2009.

13. This was a central topic of reflection on the past and future of anthropology of science and technology in the "20th Diana Forsythe Prize Celebration and Discussion," presented at the Annual Meeting of the American Anthropological Association, Vancouver, BC, November 20–24, 2019. Contributors included Heather Paxson, Sareeta Amrute, Eben Kirksey, Gabriella Coleman, Rayna Rapp, Lucy Suchman, Stefan Helmreich, Marcia Inhorn, Rene Almeling, Emily Martin, Jan English-Lueck, and David Hess.

14. I particularly enjoy thinking with a few recent books, here, including Geoffrey C. Bowker, Stefan Timmermans, Adele E. Clark, and Ellen Balka, eds., *Boundary Objects and Beyond: Working with Leigh Star* (Cambridge, MA: The MIT Press, 2015); Maria Puig de La Bellacasa, *Matters of Care: Speculative Ethics in More than Human Worlds* (Minneapolis: University of Minnesota Press, 2017); and Max Liborion, *Pollution Is Colonialism* (Durham, NC: Duke University, 2021).

METHODOLOGICAL APPENDIX

1. Diana Forsythe, "'It's Just a Matter of Common Sense': Ethnography as Invisible Work," *Computer Supported Cooperative Work* 8, no. 1–2 (1999): 127–145.

2. Two excellent texts that detail and analyze responsibility, accountability, and obligation in empirical contexts are Max Liborion, *Pollution Is Colonialism* (Durham, NC: Duke University, 2021); and Jessica M. Smith, *Extracting Accountability: Engineers and Corporate Social Responsibility* (Cambridge, MA: MIT Press, 2021). I recommend both very highly.

3. Key discussions of representation in cultural anthropology include Clifford Geertz, *The Interpretation of Cultures* (New York: Basic Books, 1973); and James Clifford and George Marcus, eds., *Writing Culture: The Poetics and Politics of Ethnography* (Berkeley: University of California Press), 1986. On the politics and limits of representation, see Audra Simpson, *Mohawk interruptus: Political Life across the Borders of Settler States* (Durham, NC: Duke University Press, 2014).

INDEX